Control Systems
Theory and Implementation

Control Systems
Theory and Implementation

Sisil Kumarawadu

Alpha Science International Ltd.
Oxford, U.K.

Control Systems
Theory and Implementation
212 pgs. | 78 figs. | 10 tbls.

Sisil Kumarawadu
Department of Electrical Engineering
University of Moratuwa
Moratuwa, Sri Lanka

ALPHA SCIENCE INTERNATIONAL LTD.

7200 The Quorum, Oxford Business Park North
Garsington Road, Oxford OX4 2JZ, U.K.

www.alphasci.com

Printed from the camera-ready copy provided by the Author.

ISBN 978-1-84265-605-1

Printed in India

To my
beloved father late Vincent Kumarawadu
and mother Yasawathi Kumarawadu

Preface

This book contains a comprehensive coverage of mathematical modeling of dynamical systems, analog and digital control principles, controller design and analysis, commercial microcontrollers/DSPs for control applications, and implementation of digital control systems using commercial processors. Theoretical contents of the book are presented as much practically oriented as possible. Heavy emphasis has been paid on the practical aspects and implementation of control systems. Digital signal processing is discussed with an explicit emphasis on realtime control applications. Control engineering is one of the broadest sub-disciplines of Engineering that can not be covered in a single book. Too much of content in the book often makes it difficult for undergraduate students and beginners to figure out which of the contents the most relevant. This book starts with the basic fundamentals, modeling of dynamical systems, discusses analog and digital control theories, and practical implementation using microprocessor-based systems. The contents cover typical syllabi of a control systems undergraduate course and postgraduate level taught courses and hence a compact yet comprehensive textbook on control systems for the budding practicing engineers.

Chapter 1 lays foundation for the reader to better understand the following chapters of the book. This chapter starts with an introduction to closed-loop control systems in terms of both hardware and system representation view points. There is a brief comparison between analog control and digital control from the points of view of theoretical as well as implementation aspects. Mathematical modeling of dynamical systems has been included. Essence of transforms in the analysis and design of dynamical systems is highlighted. Mapping between transforms used in analog and digital domains are also discussed. Concepts of system stability are discussed in the time domain, analog, and digital domains. Step response of a closed-loop system is discussed via formative mathematical analyzes. Time domain design specifications are introduced.

Chapter 2 covers the important classical graphical methods used in control theory. Root locus method is first discussed with the details of construction, reshaping, and overall design procedure outline. PID controller is detailed as one of most popular cascade compensators. The importance of frequency domain analysis and its various aspects have been included. Relation between time-domain and frequency domain specifications is discussed. Frequency domain analysis and design is addressed via reshaping the frequency response curves with special emphasis on Bode plots.

The important topics of state-space methods are covered in Chapter 3. Comparison between state-variable methods and classical methods is drawn throughout the chapter. The important topic of stability is looked at through a comprehensive mathematical analysis of the state equations.

The concepts are further elaborated by the use of some interesting practical examples. Chapter 3 also covers the concepts of controllability and observability. State-feedback control and state-feedback with integral control are discussed together with design examples.

Chapter 4 covers the important issues related to digital control theory. It covers the basic mathematics of discrete systems such as unit pulse response, difference equations, z-transform, discrete transfer functions, graphical methods and frequency response, and the methods of mapping from s to z domains. It compares and contrasts the important topic of design of digital control systems. This chapter also presents the state-space methods in the discrete-time domain.

Chapter 5 presents the fundamentals of DSPs with special emphasis on the commercial processors available for control applications and discusses general guidelines for selecting DSPs for specific control applications. Issues of sampling rate, range and round off errors associated with digital computing are discussed. DSP architectures are compared with general purpose processors. Different DSPs options are discussed in terms of arithmetic and hardware architectures, on-chip hardware and software resources. Their relative importance is discussed giving an in-depth coverage on the specific roles that each of the architectural features play in the implementation of realtime digital control systems. Software and hardware support tools for commercial DSPs are discussed with examples. Examples for practical digital implementation of variety of control algorithms and systems have also been included.

The topics nonlinear systems and intelligent control are discussed in Chapter 6. This chapter starts with an introduction to nonlinear systems. The topic of linearization of variety of nonlinear systems is then discussed giving examples. Lyapunov-based stability analysis methods are discussed in detail as the most general method for the determination of stability of nonlinear and/or time-varying systems. Rigid robot systems such as industrial robot manipulators are taken as a case study as a good practical example for complex nonlinear systems. The topic of robot control is discussed starting from fully model-based control to neural network-based online adaptive control. The concept of combined controller-observer design is discussed in detail. Design so that the nonlinearities are completely estimated online by an adaptive neural module is presented. Stability issues are addressed using formative mathematical analysis based on a Lyapunov approach. This chapter also presents fuzzy logic control (FLC) and describes two interesting design examples of FLCs.

Sisil Kumarawadu

Contents

Preface vii
List of Figures xiii
List of Tables xvii

Chapter

1 Introduction to Control Systems   **1**

1.1 Background 1
1.2 Open-loop Versus Closed-loop Control 2
1.3 Essence of Transforms in Control Theory 3
    1.3.1 An nth-order System 3
    1.3.2 Laplace Transform 4
    1.3.3 z-Transform 5
1.4 Digital Control Versus Analog Control 5
    1.4.1 Analog Versus Digital Implementation 7
    1.4.2 Analog Versus Digital System Representation 9
1.5 Continuous Control Versus Discrete Control and PLCs 11
1.6 Mathematical Modeling of Systems 12
    1.6.1 Modeling of Electrical Systems 12
    1.6.2 Modeling of Mechanical Systems 13
    1.6.3 Modeling of Electrical-mechanical Systems 15
1.7 Concepts of Stability 16
1.8 Stability and the Roots of the Characteristic Equation 17
    1.8.1 Pole-zero Cancelation 21
    1.8.2 Routh's Stability Criterion 22
1.9 Step-response and Time-domain Specifications 26
    1.9.1 Undamped Natural Frequency and the Damping Ratio 27
    1.9.2 Transient Response Specifications 27
    1.9.3 Steady State Error 29

2 Graphical Methods in Control Theory   **30**

2.1 Introduction 30
    2.1.1 Open-loop Transfer Function 31
2.2 Root Locus Method 31
    2.2.1 Construction of the Root Locus 32
    2.2.2 Reshaping the Root Locus– effects of Addition of Poles 39
    2.2.3 Reshaping the Root Locus– effects of Addition of Zeros 40
    2.2.4 Design Procedure Outline 41
2.3 PID Controller (PID Compensator) 42
2.4 Implementation of the PID Controller 45
    2.4.1 Ziegler-Nichols Method of Tuning PID Controller . . 48
2.5 Frequency-Domain Analysis 48

		2.5.1	Closed-loop Frequency Transfer Function	49
		2.5.2	Frequency-domain Specifications	50
		2.5.3	Reshaping the Frequency Response Curves	51
		2.5.4	Nyquist Stability Criterion	52
		2.5.5	Bode Plots	53
		2.5.6	Nyquist Plots	57

3 State-Space Methods — **60**

3.1	Introduction	60
3.2	State Variable Description	61
3.3	Solution of the State Equation	67
	3.3.1 State-transition Matrix	69
	3.3.2 Characteristic Equation and the Eigenvalues	71
	3.3.3 Stability and the Eigenvalues	71
3.4	Controllability and Observability	74
	3.4.1 Controllability	75
	3.4.2 Testing for Controllability	76
	3.4.3 Observability	79
	3.4.4 Testing for Observability	81
3.5	State Feedback Control	83
	3.5.1 State Feedback with Integral Control	84
3.6	Canonical Forms	89

4 Digital Control Theory — **92**

4.1	Background	92
4.2	Mathematical Methods of Discrete Systems	92
	4.2.1 A Sampled Signal	92
	4.2.2 Sampling and Data Hold	93
	4.2.3 The z-transform	95
	4.2.4 Properties of the z-transform	96
	4.2.5 Inverse z-transform	97
	4.2.6 Obtaining the z-transform when the Function in s is Known	99
4.3	Discrete Time Transfer Function	102
	4.3.1 Stability and Pole Locations	103
	4.3.2 Modified Routh's Criterion	104
4.4	Design of Digital Control Systems	108
	4.4.1 Root Locus Method	109
	4.4.2 Bode Plots	112
	4.4.3 Digital PID Control	114
	4.4.4 State Variable Methods	115
	4.4.5 Controllability and Observability	119
	4.4.6 State-feedback Control	120

5 DSPs in Control Systems — **122**

5.1	Background	122
	5.1.1 Selection of the Sampling Rate	124
	5.1.2 Range and Round-off Error	125

      5.1.3   Fixed-point Versus Floating-point Processors 126

  5.2   Single-chip DSPs 128

      5.2.1   Programming, Device Evaluation, and Debugging . . 132

      5.2.2   DSP-based Chip Sets and Control Boards 135

  5.3   Digital Implementation of Control Systems 137

      5.3.1   PID Controllers 137

      5.3.2 n^{th} Order Digital Controllers 137

      5.3.3   Motor Control 138

      5.3.4   Robot Control 142

      5.3.5   Active Power Factor Correction 145

6 Intelligent Control 148

  6.1   Linear Versus Nonlinear Systems 148

  6.2   Linearized Systems 149

  6.3   Lyapunov-based Stability Analysis of Systems 156

      6.3.1   Mathematical Background 156

      6.3.2   Stability in the Sense of Lyapunov 159

      6.3.3   Stability Analysis of Linear Systems 160

  6.4   Robot Control 162

      6.4.1   Feedback Linearizing Control 163

      6.4.2   Nonlinear Controller-observer Schemes 164

  6.5   Neurocontrol 167

      6.5.1   Radial Basis Function (RBF) NNs 168

      6.5.2   Multi-layer Perceptron (MLP) NNs 169

      6.5.3   Identification-based Indirect Control 169

      6.5.4   Direct Closed-loop Neurocontrol 171

      6.5.5   Implementation 176

  6.6   Fuzzy Logic Control (FLC) 178

      6.6.1   The Three-step Process of Generating FLCs 179

      6.6.2   Design Examples 181

  6.7   Neuro-fuzzy Systems 186

References 191

List of Figures

1.1   Block diagram of the standard error feedback control system. . .   2

1.2   Block diagram of the standard open-loop (feed-forward) control system.   3

1.3   Block diagram of the standard digital feedback control system. .   6

1.4   An operational amplifier circuit realization of the PD controller.   8

1.5   OOPic-R servo controller board that controls RC servos and small DC motors. Courtesy of Microchip Technology Inc.   9

1.6   Block diagram of an analog control system (Laplace domain representation).   10

1.7   Block diagram of a digital control system (z-domain representation).   11

1.8   Series RLC circuit.   12

1.9   The mass-spring-damper system.   14

1.10   DC motor electrical equivalent circuit.   15

1.11   Block diagram representation of the DC motor system.   16

1.12   Block diagram of an error feedback control system.   17

1.13   Purely exponential component.   20

1.14   Oscillatory components: (a) $\alpha_k = 0$, (b) $\alpha_k < 0$, (c) $\alpha_k > 0$. . . .   21

1.15   Stable and unstable regions in the s-plane.   22

1.16   System for example 1.3.   25

1.17   Transients of the unit step response of a second-order system. .   28

2.1   An analog position control system.   32

2.2   Variation of the roots with K in the s-plane.   34

2.3   The asymptotes.   38

2.4   The root locus.   39

2.5   Root locus produced by Matlab function, rlocus(sys) where sys = tf([1], [1 2 2 0]).   40

2.6   Root locus for $G(s)H(s) = K/s(s + a)$ where $a > 0$.   41

2.7   Root locus for $G(s)H(s) = K/s(s + a)(s + b)$ where $a,\ b > 0$. .   42

2.8   Root locus for $G(s)H(s) = K(s + b)/s(s + a)$ where $a,\ b > 0$. .   43

2.9   An error feedback control system.   44

2.10   An Operational amplifier circuit realization of the PI controller.   45

2.11   Trapezoidal integration.   46

2.12   Typical gain-phase curves of a feedback control system.   50

2.13   Effect of adding a pole at $s = -1/b$: (1) $b = 5$, (2) $b = 1$, (3) $b = 0.5$, (4) $b = 0$.   52

2.14   Bode plot of a stable system: $GM, PM > 0$, $\omega_c > \omega_\phi$.   53

2.15   Bode plot of an unstable system: $GM, PM < 0$, $\omega_c < \omega_\phi$.   54

2.16 Bode plot produced by the Matlab command, bode(50, [1 9 30 40]). 56

2.17 Bode plot produced by the Matlab command, bode(40, [1 7 10 0 0]). 58

3.1 Dynamical system with its internal states, inputs and outputs . 61

3.2 Equivalent circuit of a separately exited dc motor 63

3.3 DC motor speed control system 65

3.4 State space trajectory 75

3.5 Wheatstone bridge circuit 77

3.6 State feedback control. 83

3.7 State feedback control with an observer. 84

3.8 State feedback with integral control. 85

3.9 Inverted pendulum. 87

4.1 (a) Unit pulse function, (b) a continuous signal, (c) a sampler, and (d) the sampled signal. 93

4.2 zero-order hold. 94

4.3 Step response of the system: Example 4.3. 103

4.4 Mapping from s-plane to z-plane. 104

4.5 System for Example 4.5. 106

4.6 Sampled data system for Example 4.6. 109

4.7 Matlab plot for Example 4.6. 111

4.8 z-grid plot for Example 4.6. 113

4.9 State feedback control. 121

5.1 ADSP 2102 based digital controller. Courtesy of Analog Devices, Inc. 124

5.2 DSP56000 family DSP core diagram. Courtesy of Motorola, Inc. 129

5.3 Data registers and MAC/ALU of DSP56000 core. X0, X1, Y0, Y1 are data input registers and A, B are accumulator registers. 130

5.4 Development tools for DSP. 135

5.5 DSP-based 16-axis motion control board. Courtesy of ADLINK Technology Inc. 136

5.6 2nd order biquad filter 138

5.7 AC servomotor control system. 140

5.8 DSP-based AC servomotor control system. 141

5.9 F.A.A.K mobile robot system. Courtesy of Control Systems Engineering Group, Fern Universität in Hagen 144

5.10 Implementation of PFC. Courtesy of Microchip Technologies Inc. 147

6.1 Kinematic model of the robotic head. 152

6.2 Binocular head configuration parameters (q_p, q_t, q_v) with respect to the world coordinate frame $\{\hat{X}_W, \hat{Y}_W, \hat{Z}_W\}$. B is the baseline distance. 153

6.3 The Katana robotic arm. Courtesy of Neuronics AG 162

6.4 Block diagram of the autonomous vehicular system. 170

6.5 Block diagram of manual driving set up during data collection. . 171

6.6 Joint angle trajectories-rigid robot arm. 177

6.7 Neural network functional estimates. 177

6.8 The relative membership in a fuzzy set. 179

6.9 Centroidal defuzzification. 181

6.10 The control system block diagram for autonomous boat. 182

6.11   Membership functions of the position variables.   182
6.12   Membership functions for the heading variable. Desired value is
          180°.   183
6.13   Membership functions for rudder angle.   183
6.14   An example on how maneuvers are defined differently depending
          on the distance to the target point.   184
6.15   The control system block diagram for cyclical leg motion.   185
6.16   ANFIS architecture.   187

List of Tables

1.1 Routh's array 23
1.2 Routh's array for $1 + G(s)H(s) = s^3 + 4s^2 + 8s + 12$ 23
1.3 Routh's array for $1 + G(s)H(s) = s^4 + s^3 + 2s^2 + 2s + 5$ 24
1.4 Routh's array for $1 + G(s)H(s) = s^5 + 2s^4 + 3s^3 + 4s^2 + 7s + 5$ 24
1.5 Routh's array for Example 1.2 25
2.1 Roots of $s(s + 2) + K = 0$ 33
4.1 z-transform of some common functions 96
5.1 Some popular single-chip DSPs 133
5.2 Some popular single-chip DSPs, contd. 134
6.1 Fuzzy PD control rule base for cyclic leg motion 186

Chapter 1

Introduction to Control Systems

1.1 Background

To start with, it is important to understand the concept of feedback or closed-loop control. Consider writing *University of Moratuwa* on the white board with a marker pen attached to someone's fingers. Here, there is a perfect feedback control system in that eyes are used to capture the images of what is actually being written, alignments, spacing between the words, clarity etc. These images are encoded and sent to the brain for processing. Brain, knowing what is expected to be written and other specifications such as alignments and clarity, can now compare what is expected (reference) and what is actually happening (measured output based on visual sensor information) and control the movements of the muscles of the arm and hand in a way the difference between the reference and the output is always kept minimum. Difference between the reference and the output is usually referred to as the *error* in control systems terminology.

Above is an example for a feedback control system, which is completely biological. Automatic control is referred to as control of dynamical systems without a direct human involvement during the operation. For instance, by the use of today's technology, one can automate the aforementioned totally biological control system by replacing the eyes by video cameras, brain by a digital computer, hand by a robot arm, muscles by electric motors, and hand by a suitable robot end-effector. Marker pen is just a tool and may remain the same in both the systems. One may define the desired trajectory of the pen tip in terms of $(x,\ y)$ coordinates in a Cartesian frame that is fixed on the white board. Machine vision or computer vision is a rapidly growing technology in which digital computer is used to process and analyze the images captured by cameras. Image processing followed by some coordinate transformations can help recover the actual trajectory of the pen tip. The control algorithm, which is a software program running on another digital computer should be capable

of producing control commands to the electric motors in a way the error
is always regulated at zero.

Fig. 1.1 indicates a standard error feedback control system configuration. The plant or the controlled system is the system that is controlled.
The feedback element is typically a sensor that feeds the plant output back
to be used by the controller. Often, the most challenging task of the designer is to determine the structure of the controller, which is driven by
the difference between the reference input and the fed-back output signals.
Understandably, the reference input is the desired output.

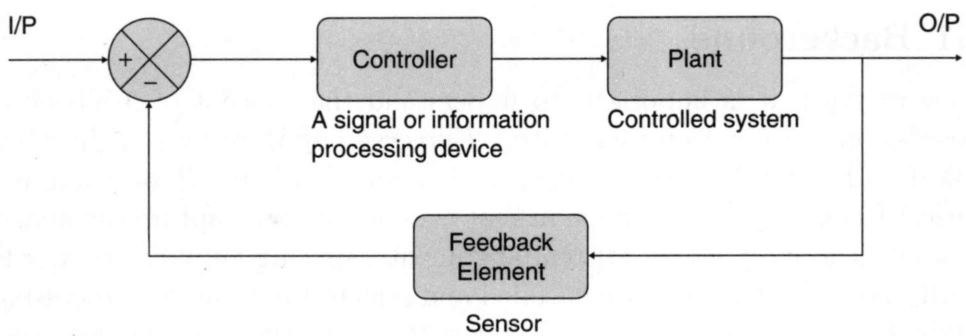

Figure 1.1: Block diagram of the standard error feedback control system.

1.2 Open-loop Versus Closed-loop Control

Consider writing *University of Moratuwa* on the white board with a marker
pen, but this time with the eyes closed. Even though the past experience
may enable doing a reasonable job, if tried to write it a bit faster may
result in alignment, spacing issues etcetera producing a piece of writing
that is unacceptable. The reason for this is the absence of feedback to be
compared with the reference input or the desired way of writing. There
is no way to correct the movement of the pen as the brain can not know
what is exactly happening.

Another example is the fact that a deaf can not speak properly even
though he may have gone deaf after he learned how to speak. In the error
feedback control point of view, what he actually wants to speak represents
the reference input and what is actually being spoken is the sensed output.
Output quantity has no influence in the input quantity (and hence in the
controller) and hence there is no feedback in the control system. Fig. 1.2
indicates a standard open-loop control system configuration.

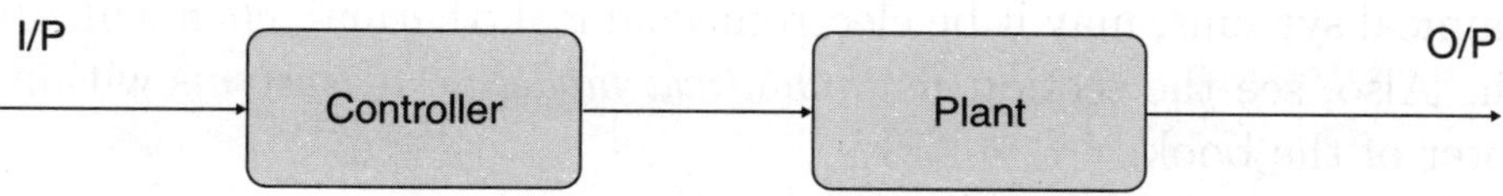

Figure 1.2: Block diagram of the standard open-loop (feed-forward) control system.

A common automatic control example for open-loop control is domestic bread toaster. The setting of the darkness knob or timer represents the reference input, and the degree of darkness or crispness of the toasted bread is the output. If the degree of darkness is not satisfactory, may be because the type of bread is different, there is no way to automatically alter the length of time the heat is supplied. The sensors that can be considered here may be touch sensors or miniature color camera or both. But incorporating such sensors and accessories in high temperature environment is not an practical option due to added complexity and the cost. Furthermore, high precision control is not an issue of concern in this application. In general, the reasons for opting for the option of open-loop control may be either high-precision control is not required or sensing is practically impossible or financially unjustifiable or both.

1.3 Essence of Transforms in Control Theory

1.3.1 An nth-order System

Applied mathematics is an essential tool in the studies and theoretical analyzes of control systems. If the mathematical models of the systems are available, the designer, through analyzing them, can arrive at reasonably predictable and reliable designs without depending heavily on thorny repetitive real-world experimentations and extensive computer simulations. Dynamical systems can be mathematically modeled by differential equations. These equations generally involve derivatives and integrals of the independent variables with respect to the dependent variable. Consider a moving ground, aerial, or underwater vehicle. Equations of its translational motion can be modeled by the use of Newton's 2nd law, $F = m\ddot{x}$, where F is the total external force in the x-direction, m is the mass, and $\ddot{x}$ is the acceleration in the x-direction. Once substituted for F, one ends up with a differential equation. Consider an electric circuit that in general consists of resistors, capacitors, and inductors. Writing Kirchoff's laws give the governing differential equations of the circuit. Same is valid for other dynamical systems, may it be electromechanical, thermal, chemical, and so forth. Also, see the section *mathematical modeling of systems* within this chapter of the book.

In general, an nth order differential equation is written as

$$a_n \frac{d^n y(t)}{dt^n} + a_{n-1} \frac{d^{n-1} y(t)}{dt^n} + \cdots + a_1 \frac{d^n y(t)}{dt^n} + a_0 y(t) = u(t) \qquad (1.1)$$

where $u(t)$ is the independent variable and $y(t)$ is the dependent variable or the unknown that is to be determined by solving the differential equation. If a system can be mathematically modeled by an nth order differential equation, we refer to that system as an nth-*order system*.

Equation (1.1) is called a *linear ordinary differential equation* if its coefficients $a_n, a_{n-1}, \cdots, a_0$ are not functions of $y(t)$. If the coefficients are constants, the differential equation is known as time-invariant. For instance, assume that the resistance is time varying in an electric R, L, C circuit. Then, the resulting differential equation

$$R(t)i(t) + L\frac{di(t)}{dt} + \frac{1}{C}\int i(t)dt = e(t) \qquad (1.2)$$

where $e(t)$ is the applied voltage, will be time-variant even though the co-efficients may not be functions of the dependent variable, $i(t)$, the current in the network. Laplace transform method is extensively used to solve ordinary constant coefficient (linear-time-invariant or LTI) differential equations.

1.3.2 Laplace Transform

The Laplace transform is an efficient mathematical tool used to solve linear ordinary differential equations. The Laplace transform transforms the problem of differential equations in the time domain into algebraic equations in the Laplace domain. Laplace operator, s, is a complex variable: i.e., $s = \sigma + j\omega$. Manipulating algebraic equations of s by algebraic methods such as factorization, partial fractions, etc is comparatively much easier. Solutions obtained in the Laplace domain or s-domain are taken back to the time domain by performing inverse Laplace transformation.

Laplace transform of a time signal, $f(t)$, is formally denoted by $\mathcal{L}\{f(t)\} = F(s)$ and is defined as

$$F(s) = \int_0^\infty f(t)e^{-st}dt. \qquad (1.3)$$

Since the integration is evaluated from $t = 0$, all information contained in $f(t)$ prior to $t = 0$ is ignored. This is very much okay with *causal* systems where response do not precede excitation. In time domain studies, time reference is often chosen to be $t = 0$. For instance, to check the step response of a system, step change in excitation can often be made at $t = 0$.

If $F(s)$ is known, the time signal can be obtain by taking the inverse Laplace transform of $F(s)$ as $f(t) = \mathcal{L}^{-1}\{F(s)\}$. Therefore, $f(t) \leftrightarrow F(s)$ is called a Laplace transform pair. Study of Laplace transform is an essential integral part of an applied mathematics course taught in an engineering undergraduate program and is omitted in this book.

1.3.3 z-Transform

z-transform is the discrete-domain counterpart of Laplace transform. Difference equations is the discrete-domain counterpart of differential equations and the z-transform is an efficient mathematical tool used to solve linear ordinary difference equations. Quite analogous to Laplace transform, z-transform transforms the problem of difference equations in the time domain into algebraic equations in the z- domain. Solutions obtained in the z-domain are taken back to the time domain by performing inverse z-transform.

z-operator is the discrete counterpart of the Laplace operator, s, and an explicit relationship exists between the two operators. This will be discussed shortly within this chapter. Actually, z-transform is the Laplace transform of a sampled signal.

1.4 Digital Control Versus Analog Control

Traditionally, control systems have been designed and analyzed using analog techniques such as differential equations and Laplace transform-based methods. Furthermore, the controllers have been implemented using analog hardware components, such as resisters, capacitors, and operational amplifiers whenever dealing with electrical signals. In totally mechanical systems, springs, dampers, and mass elements were used. Today, because of explosive growth of digital technology in terms of speed, accuracy, efficiency coupled with ever increasing affordability, controllers are typically implemented as programable digital hardware such as PLCs that will be discussed shortly or software programs on digital computers.

Real world signals are inherently analog. As digital computers can not work with analog signals, these signals need to be sampled and converted to digital form by the use of an analog-to-digital converter (ADC). Likewise, the digital output signals from a digital computer should be converted back to analog form by the use of a digital-to-analog converter (DAC). When the analog computer in Fig. 1.1 is replaced by a digital computer, ADCs, and DACs, such a system is called sampled-data system or a digital control system. Fig. 1.3 depicts a standard digital feedback control system configuration. Note that the controlled system and sensors that are inherently analog remain the same.

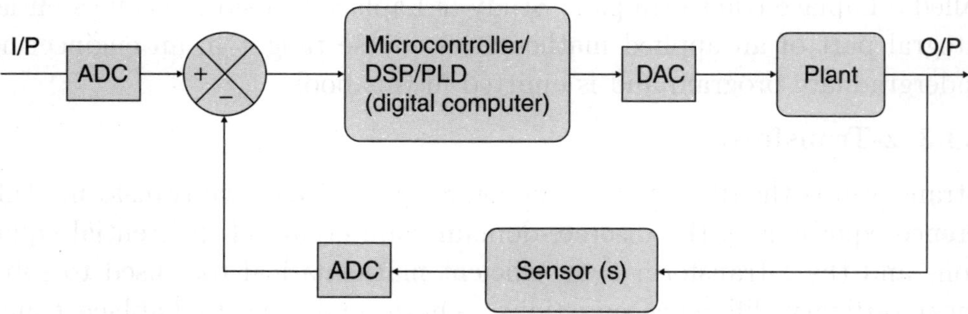

Figure 1.3: Block diagram of the standard digital feedback control system.

Microcontroller is a microprocessor that is specially designed for control applications. Microcontrollers often have hardware architecture aspects significantly different from a general purpose microprocessor. Digital signal processors (DSPs) are available off-the-shelf today for numerous different applications that include control applications (induction motor control, high-precision servo motor control, industrial robot control, engine control, active suspension control of automobiles, and many more), communication, speech recognition, vision, and Hi-fi audio encoding and decoding. DSPs for control applications are specially designed microcomputer systems with special hardware architectures and other on-chip hardware and software resources to best serve the challenging task of implementing complex control systems realtime.

From hardware point of view, even though the term *controller* may refer to the digital computer and microcontroller such as DSP on which the control algorithm is programmed, term *controller* may also mean the control algorithm itself. Control algorithm or control law is the algorithm that generates control commands to the plant in a way the error is always regulated at zero. Within the closed-loop of a feedback system, a DSP functions as an information processing device. It processes the information that it receives and outputs signals by the use of software programs.

1.4.1 Analog Versus Digital Implementation

When it comes to error feedback control, arguably, the simplest controller that one can ever think about is a controller that is a simple amplifier with a constant gain, K. This type of control action is formally known as *proportional control*, since the control command that the controller produces is simply proportional to the input to the controller. In error feedback control, for instance, control command $u(t) = Ke(t)$ where $e(t)$ is the error. Controller produces a larger control command if the error is large, and if $K > 0$, with a polarity being the same as that of the instantaneous error. Even though error may be brought to zero faster by selecting a large value for K, this will give rise to large overshoot and settling times, and hence poor transient performance. To overcome this, one may also include a derivative terms as follows

$$u(t) = K_P e(t) + K_D \frac{de(t)}{dt} \tag{1.4}$$

or in the Laplace domain

$$\frac{U(s)}{E(s)} = G_c(s) = K_P + K_D s \tag{1.5}$$

where $G_c(s)$ is formally known as transfer function of the controller with $u(t) \leftrightarrow U(s)$ and $e(t) \leftrightarrow E(s)$ being Laplace transform pairs. This is the popular PD (Proportional + Derivative) control law that is heavily adopted in the industry. As derivative control action in PD control law is sensitive to the rate of change of error that is large during the transients, now, both transient and steady state responses may be improved by fine tuning K_P and K_D.

This controller can be implemented either using analog hardware or as a program on a digital computer. An operational amplifier circuit realization of the PD controller is given in Fig. 1.4. Note that the second half is an inverting circuit. The transfer function of the circuit is

$$\frac{U(s)}{E(s)} = G_c(s) = \frac{R_2}{R_1} + R_2 C_1 s. \tag{1.6}$$

In control theory, the term *transfer function* is defined as the ratio of Laplace transform of output to Laplace transform of input. Comparing (1.6) with (1.5), we get

$$K_P = R_2/R_1 \qquad K_D = R_2 C_1. \tag{1.7}$$

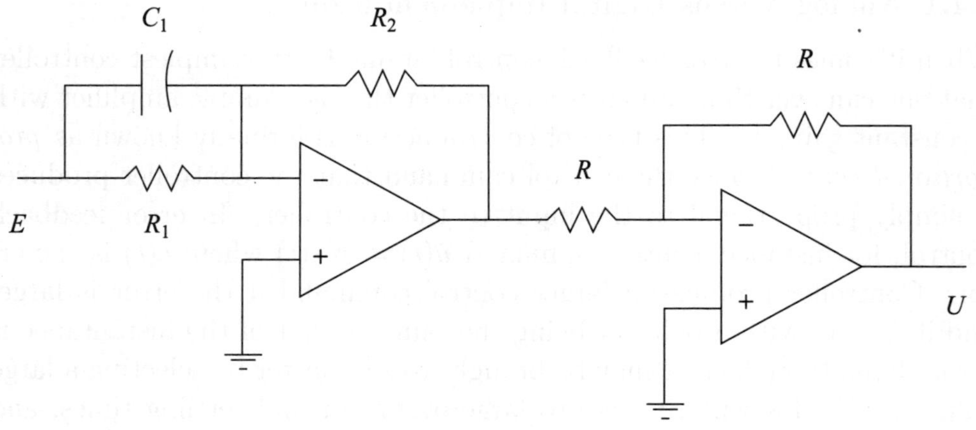

Figure 1.4: An operational amplifier circuit realization of the PD controller.

Now, the gains, K_P and K_D, can be adjusted by suitably selecting the values for R_1, R_2, and C_1.

While the controller in (1.5) has traditionally been implemented with analog hardware that work with continuous-time analog signals, the more modern trend is towards digital control. Fig. 1.3 illustrates a general digital control system that we can view as a control system with control algorithm being PD control, if the controller block in the block diagram includes the discrete version of the PD controller. Quite in contrast to the analog implementation, now, the control algorithm is a program on a microcontroller and the equation might take the simple form

$$u(k) = K_P e(k) + K_D \frac{e(k) - e(k-1)}{T} \tag{1.8}$$

where $u(k)$ is the control command to the plant at time k, $e(k)$ the error at time k, $e(k-1)$ the error at previous sample, and T is the constant sampling period. K_P and K_D are gain values to be chosen by the designer. The control equation, (1.8), has a transfer function form that follows from z-transforming both sides of it as follows

$$\frac{U(z)}{E(z)} = G_c(z) = K_P + K_D \frac{z-1}{Tz}. \tag{1.9}$$

The z-transform theory and related issues will be detailed in chapter 4.

The microcontroller given in Fig. 1.5 is programmable in Object Oriented Basic, C or Java syntax styles. These microcontrollers, among

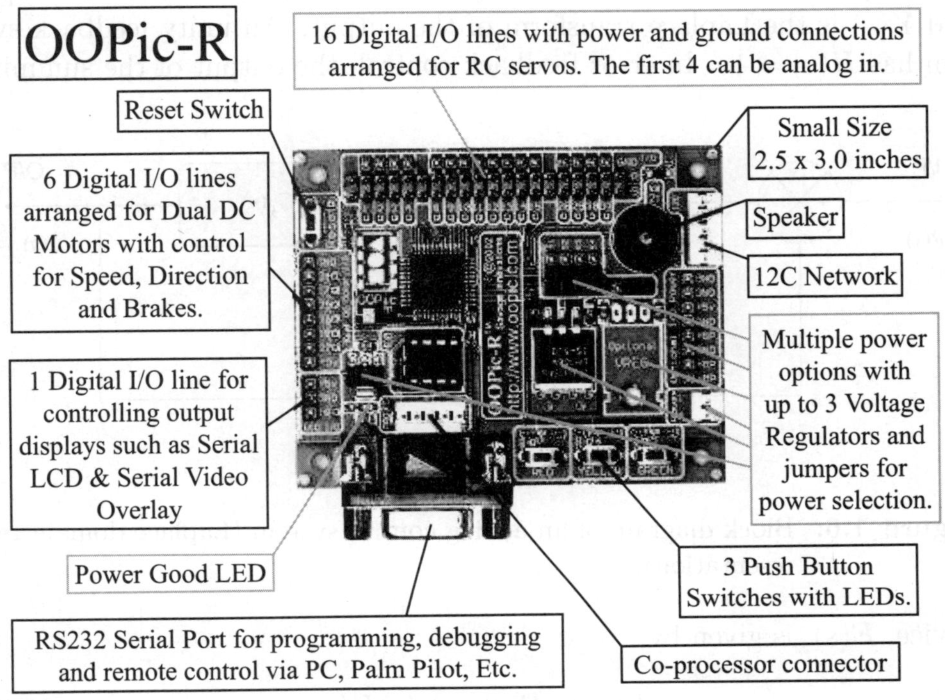

Figure 1.5: OOPic-R servo controller board that controls RC servos and small DC motors. Courtesy of Microchip Technology Inc.

other things, implement equations such as control laws. Real world control problems may be computationally expensive with control equation of more complex forms. DSPs-based implementation is often preferred to meet the challenges imposed by real time requirements in implementing complex control systems. Commercial DSP-based control solutions come in variety of forms such as single-chip DSPs, DSP chip sets, DSPs-based control boards, etc. These important topics will be discussed in chapter 5.

1.4.2 Analog Versus Digital System Representation

Fig. 1.6 shows a block diagram representation of a typical analog error feedback control system from mathematical point of view. The basic components of the loop, viz., the controller, the plant, and the feedback blocks, are mathematically represented in the Laplace domain, respectively, by the transfer functions, $G_C(s)$, $G_P(s)$, and $H(s)$ of each block. Controllers are algorithms implemented by the use of analog hardware to obtain desired control performance. $R(s)$ is the Laplace transform of the reference input and $Y(s)$ is the Laplace transform of the output. An unity feedback system has $H(s) = 1$. In error feedback control, the output of the summing

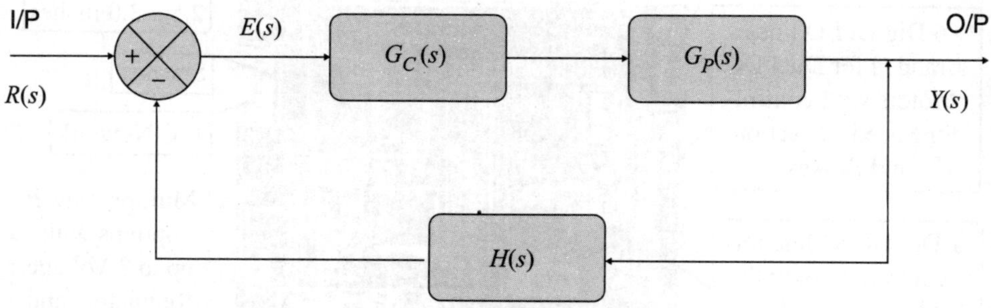

Figure 1.6: Block diagram of an analog control system (Laplace domain representation).

device, $E(s)$, is given by

$$E(s) = R(s) - Y(s)H(s). \tag{1.10}$$

Also

$$Y(s) = E(s)G_C(s)G_P(s). \tag{1.11}$$

Once the two preceding equations are combined, we get

$$\frac{Y(s)}{R(s)} = M(s) = \frac{G_C(s)G_P(s)}{1 + G_C(s)G_P(s)H(s)}. \tag{1.12}$$

$M(s)$ is called the *closed-loop transfer function.*

Fig. 1.7 shows a block diagram representation of a typical digital control system (error feedback). The controller, $G_C(z)$, is usually the control algorithm programmed on a microcontroller or a DSP. DACs and ADCs may or may not be considered as separate elements. These issues will be discussed in chapter 4. In Fig. 1.7, they are assumed to be part of the blocks, $H(z)$ and $G_C(z)$. Let $R(z)$ be the z-transform of the input and $Y(z)$ the z-transform of the output. Following the same lines as in the analog case, we can derive the closed-loop transfer function of the digital control system in Fig. 1.7 as

$$\frac{Y(z)}{R(z)} = M(z) = \frac{G_C(z)G_P(z)}{1 + G_C(z)G_P(z)H(z)}. \tag{1.13}$$

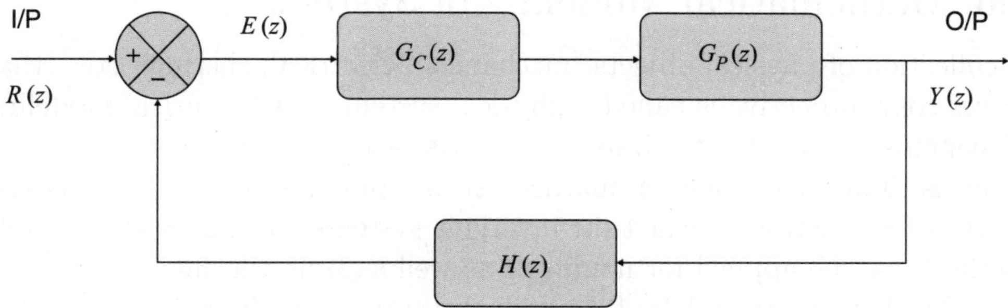

Figure 1.7: Block diagram of a digital control system (z-domain representation).

1.5 Continuous Control Versus Discrete Control and PLCs

The control systems discussed so far belong to continuous control because the objective of the control therein is to control a continuous variable such as temperature, speed or shaft position of a motor, and so forth. Discrete control that is widely implemented using programmable logic controllers (PLCs) refers to control of the order and timing of a set of discrete events. A common example is the operation of a washing machine.

In discrete control, changes are defined in advance by means of a program of instructions. The change can be to initiate an operation or terminate an operation, start a motor or stop it, open a valve or close it and so forth. The changes are executed either because the state of the system has changed or because a certain amount of time has elapsed. In the operation of the washing machine, soon after the lid is closed, an inlet valve opens to start filling the laundry tub with water. Once the laundry tub has been filled to the preset level, water inlet is closed and agitation starts. Agitation cycle lasts for a length of time set on the controls. When this time is up, the timer stops the agitation and initiates the draining of the tub. In discrete control using PLCs, *combinational logic control* is used to control event-driven changes, such as filling the laundry tub with water. Filling continues until the preset level has been sensed, which causes the inlet valve to close. *Sequential logic* is used to manage time-driven changes, such as duration of the agitation cycle.

1.6 Mathematical Modeling of Systems

A collection of physical objects (mechanical, electrical, thermal, etc.) that serves some objective is called a physical system. Mathematical modeling of systems is one of most important tasks of control systems design and analysis. Two most common mathematical representations of systems are: 1] Transfer function (linear time invariant systems only) 2] State-variable method (can be applied for nonlinear as well as time varying systems).

Mathematical models of dynamical systems are differential equations. For instance, the translational motion of a moving object under the influence of external forces can be described using Newton's 2nd law. When expressed in terms of mathematical equations, we end up with differential equations. The governing equations of an electric motor can be written referring to its electrical equivalent circuit and mechanical sub-systems. They necessarily are differential equations. A linear system is a system that can be represented using linear differential equations. If the coefficients of the differential equations are constant and do not vary with time, we have a time invariant system.

1.6.1 Modeling of Electrical Systems

Consider the RLC circuit in Fig. 1.8. Kirchoff's voltage law yields

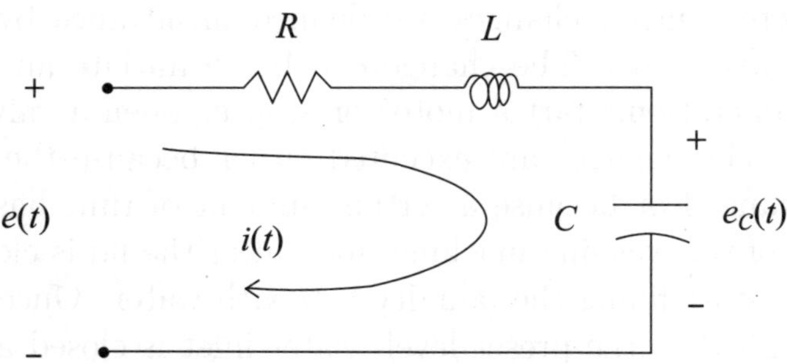

Figure 1.8: Series RLC circuit.

$$e(t) - R\, i(t) - e_c(t) = L\frac{di(t)}{dt}. \tag{1.14}$$

Current in the capacitor is

$$i(t) = C\frac{de_c(t)}{dt}. \tag{1.15}$$

Rewriting the preceding equations using matrix notation yields

$$\begin{bmatrix} \dot{e}_c(t) \\ \dot{i}(t) \end{bmatrix} = \begin{bmatrix} 0 & 1/C \\ -1/L & -R/L \end{bmatrix} \begin{bmatrix} e_c(t) \\ i(t) \end{bmatrix} + \begin{bmatrix} 0 \\ 1/L \end{bmatrix} e(t) \qquad (1.16)$$

or

$$\dot{\boldsymbol{x}} = A\boldsymbol{x} + Bu \qquad (1.17)$$

where $\dot{\boldsymbol{x}} = [e_c(t)\ i(t)]^T$ is called the state vector and $u = e(t)$ is the input.

One can obtain the following transfer functions of the system by taking the Laplace transforms of (1.14) and (1.15) and making all the initial conditions to zero

$$\frac{E_c(s)}{E(s)} = \frac{1}{1 + RCs + LCs^2} \qquad (1.18)$$

$$\frac{I(s)}{E(s)} = \frac{Cs}{1 + RCs + LCs^2}. \qquad (1.19)$$

Note that $i(t) \leftrightarrow I(s)$, $e(t) \leftrightarrow E(s)$, and $e_c(t) \leftrightarrow E_c(s)$ are Laplace transform pairs. Initial conditions are not taken into account in the transfer function representation of the systems. Selection of the input and the output in the transfer function representation of a plant often depends on the control task and the strategy how it is achieved.

1.6.2 Modeling of Mechanical Systems

Newton-Euler formulation of body dynamics is often used to model mechanical systems. For instance, the motion of vehicular or robot system can be described in various dimensions as translational, rotational, or combinations.

The motion variables that are used to describe translational motion are linear acceleration, velocity, and displacement. Newton's second law of motion states that the resultant force acting on a rigid body in a given direction is the product of the mass of the body and its acceleration in the same direction. the law can be expressed as

$$\sum \text{external forces} = M\ddot{x} \qquad (1.20)$$

where M and $\ddot{x}$, respectively, are the mass and the linear acceleration. For instance, the equations for the lateral and longitudinal motions of a vehicle can be obtained by applying the above expression in the two directions when all the external forces acting upon the vehicle are known.

The rotational motion of a body is defined as motion about a fixed axis. The extension of Newton's second law of motion states that the algebraic sum of moments or torque about a fixed axis acting on a rigid body is equal to the product of the inertia of the body about the same axis and the angular acceleration about the axis. The law can be expressed as

$$\sum \text{external torques} = I\ddot{\theta} \tag{1.21}$$

where I and $\ddot{\theta}$, respectively, are the inertia and the angular acceleration. The yaw, roll, or pitch dynamics of a vehicle, for instance, can be derived using the above equation. An in-depth analysis on modeling of real-world systems can be found in contemporary literature such as [1].

Example: Like electrical systems are modeled as RLC circuits, mechanical systems are modeled using ideal spring, damper, and mass elements. Consider the mass-spring-damper system given in Fig. 1.9. Here,

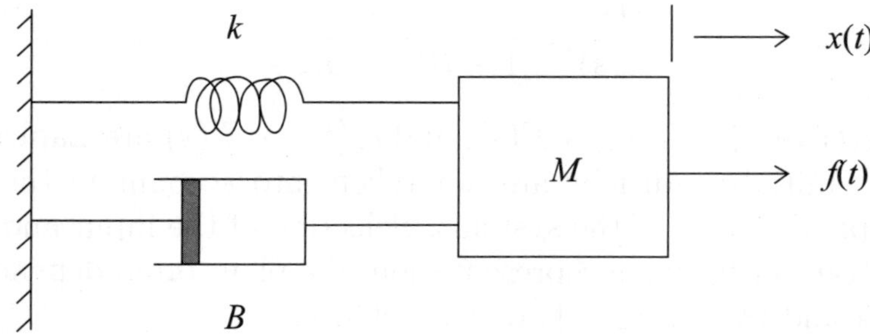

Figure 1.9: The mass-spring-damper system.

$x(t)$ is the displacement and $f(t)$ is the external force applied. For a spring in the linear range

$$\text{force} = kx(t).$$

Also

$$\text{friction force} = B\dot{x}(t).$$

Writing Newton's equation for the linear motion of the mass yields

$$M\ddot{x}(t) = -B\dot{x}(t) - kx(t) + f(t). \tag{1.22}$$

By defining the state variables as $x_1 = x$, $x_2 = \frac{dx_1}{dt}$, one gets the state equations as

$$\frac{dx_1(t)}{dt} = x_2(t) \tag{1.23}$$

$$\frac{dx_2(t)}{dt} = -\frac{k}{M}x_1(t) - \frac{B}{M}x_2(t) + \frac{1}{M}f(t). \tag{1.24}$$

In the matrix form $\dot{\boldsymbol{x}} = A\boldsymbol{x} + Bu$ with $u = f(t)$ being the external input to the mass-spring-damper system.

Taking the Laplace transform of the original equation, one can obtain the transfer function as

$$\frac{X(s)}{F(s)} = \frac{1}{Ms^2 + Bs + k} \tag{1.25}$$

where $X(s)$ and $F(s)$, respectively, are the Laplace transforms of the output and the input.

1.6.3 Modeling of Electrical-mechanical Systems

Electric motor is an electrical-mechanical system. Electrical motors play an important role in many practical control systems. DC motors are used extensively in control systems. In permanent magnet (PM) DC motors, magnetic field is produced by a permanent magnet. The electrical equivalent circuit of a PMDC motor is shown in Fig. 1.10. The subscript a

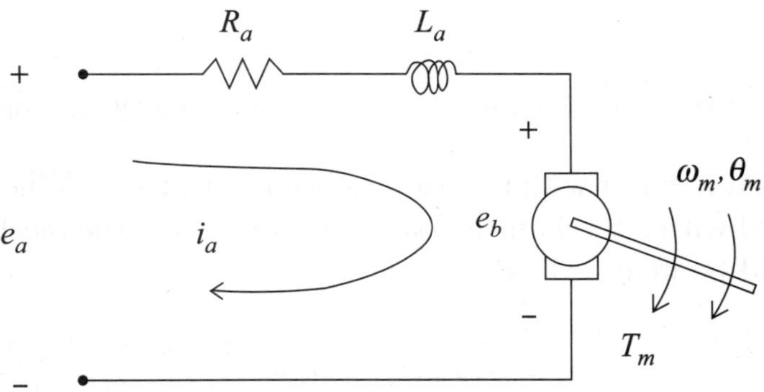

Figure 1.10: DC motor electrical equivalent circuit.

stands for armature. Motor torque is given by $T_m(t) = K_i\, i_a(t)$ with k_i being a constant for constant magnetic flux.

Following equations can be written for the electrical and mechanical sub-systems

$$\frac{di_a(t)}{dt} = \frac{1}{L_a}e_a(t) - \frac{R_a}{L_a}i_a(t) - \frac{1}{L_a}e_b(t) \tag{1.26}$$

$$T_m(t) = k_i i_a(t) \tag{1.27}$$

$$e_b(t) = k_b\frac{d\theta_m(t)}{dt} = k_b\omega_m(t) \tag{1.28}$$

$$\frac{d^2\theta_m(t)}{dt^2} = \frac{1}{J_m}T_m(t) - \frac{B_m}{J_m}\frac{d\theta_m(t)}{dt}. \tag{1.29}$$

By solving the Laplace transforms of the above equations and ignoring the initial conditions, the transfer function between motor shaft displacement and the input voltage is

$$\frac{\theta_m(s)}{E_a(s)} = \frac{k_i}{L_aJ_ms^3 + (R_aJ_m + B_mL_a)s^2 + (k_bk_i + R_aB_m)s}. \tag{1.30}$$

The block diagram representation is given in Fig. 1.11.

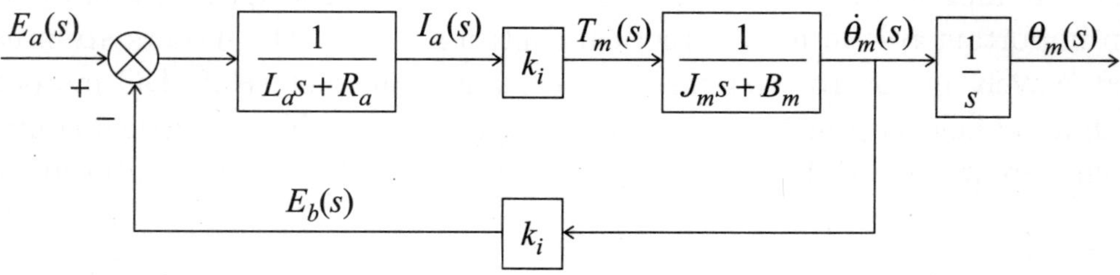

Figure 1.11: Block diagram representation of the DC motor system.

Note that a motor with no load has been considered. When the motor shaft is loaded with a mechanical load, the equation of the mechanical subsystem should be modified as

$$\frac{d^2\theta_m(t)}{dt^2} = \frac{1}{(J_m + J_l)}(T_m(t) - T_l) - \frac{(B_m + B_l)}{(J_m + J_l)}\frac{d\theta_m(t)}{dt} \tag{1.31}$$

where the subscript l stands for load.

1.7 Concepts of Stability

In general, the state or quality of being stable refers to steadiness, firmness or strength to stand without being overthrown. Assume an application of automatic vehicle following problem where the agent or the following vehicle is controlled to automatically follow a lead vehicle maintaining a safe distance. Top most concern in the automatic vehicle following controller design is to make sure that the vehicles do not crash or get out of the road. Ensuring controller performance specifications such as ride comfort (less velocity fluctuations, smooth accelerations etc) or offset (ensuring the difference between the desired inter-vehicular spacing and the actual spacing) are secondary and should only be taken into consideration after ensuring the stability. In a controller design problem to control a robot arm to perform an automatic cutting job with a cutting tool attached at the robot end effector, first task is to ensure stability to ensure that the error between the desired trajectory of the tool and the actual trajectory stays bounded so that fine tuning to improve the control accuracy is now possible. To that end, in control engineering point of view, stability refers to error being bounded without increasing indefinitely. This definition may sound rather mathematical, because in practice, safety measures will be activated before error reaches hazardous levels, let alone infinity.

Among many forms of performance specifications used in design, the most important requirement is that the system be stable. An unstable system is generally considered to be useless. The Lyapunov-based stability notions that will be discussed in chapter 6 are very general and widely used in stability proofs in nonlinear control design.

1.8 Stability and the Roots of the Characteristic Equation

In control systems, the system stability should be first ensured before the system performance is fine tuned to obtain the precision required. The stability characteristics of a linear time-invariant system can be determined from the system's characteristics equation.

Let the feedback (FB) control system be given by Fig. 1.12. System overall transfer function is

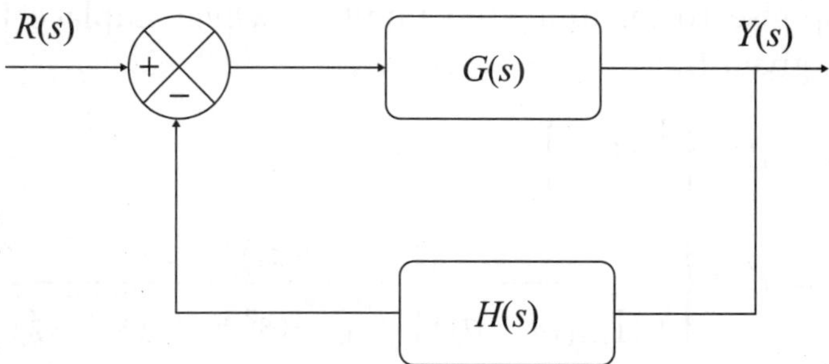

Figure 1.12: Block diagram of an error feedback control system.

$$M(s) = \frac{Y(s)}{R(s)} = \frac{G(s)}{1 + G(s)H(s)} \qquad (1.32)$$

$$= \frac{(s+z_1)(s+z_2)\cdots(s+z_n)}{(s+p_1)(s+p_2)\cdots(s+z_m)}. \tag{1.33}$$

$M(s)$ with $m > n$ is a rational function of s and hence can be decomposed into partial fractions. Also,

$\quad -z_1, -z_2, \cdots, -z_n$ are the zeros of the system or closed-loop zeros.

$\quad -p_1, -p_2, \cdots, -p_m$ are the poles of the system or closed-loop poles.

Poles of the closed-loop transfer function are the same as the roots of the characteristics equation that is defined as

$$1 + G(s)H(s) = 0. \tag{1.34}$$

Poles should either be real or complex conjugate pairs. Therefore, $M(s)$ may also be written as

$$M(s) = \frac{\prod_{i=1}^{n}(s+z_i)}{\prod_{j=1}^{r}(s+\sigma_j)\prod_{k=1}^{\frac{1}{2}(m-r)}(s^2 + 2\alpha_k s + \omega_{mk}^2)}. \tag{1.35}$$

Note in the preceding equation that in the denominator first r terms represent the real roots while the latter terms represent the complex conjugate pairs. One of the ways to study the stability issues is to consider the step response of the closed-loop system. As the Laplace transform of the output signal is

$$Y(s) = R(s)M(s) \tag{1.36}$$

the time response of the closed-loop system is

$$y(t) = \mathcal{L}^{-1}\left\{R(s)M(s)\right\}. \tag{1.37}$$

The time response to an unit step input of which Laplace transform is $R(s) = 1/s$ is given by

$$y(t) = \mathcal{L}^{-1}\left\{\frac{1}{s}M(s)\right\}$$

$$= \mathcal{L}^{-1}\left\{\frac{\prod_{i=1}^{n}(s+z_i)}{s\prod_{j=1}^{r}(s+\sigma_j)\prod_{k=1}^{\frac{1}{2}(m-r)}(s^2 + 2\alpha_k s + \omega_k^2)}\right\}. \tag{1.38}$$

Decomposing the preceding equation into partial fractions yields

$$y(t) = \mathcal{L}^{-1}\left\{\frac{A}{s} + \sum_j \frac{B_j}{s+\sigma_j} + \sum_k \frac{C_k s + D_k}{s^2 + 2\alpha_k s + \omega_k^2}\right\}$$

$$= A + \sum_j B_j e^{-\sigma_j t} + \sum_k \mathcal{L}^{-1}\left\{\frac{C_k s + D_k}{s^2 + 2\alpha_k s + \omega_k^2}\right\} \tag{1.39}$$

where A, B_j, C_k, and D_k are constants that can be evaluated by comparing the coefficients with those of the original expression. Now, consider the last term of the preceding equation

$$
\frac{C_k s + D_k}{s^2 + 2\alpha_k s + \omega_k^2} = \frac{C_k(s + \alpha_k)D_k - C_k\alpha_k}{(s + \alpha_k)^2 + \omega_k^2 - \alpha_k^2}
$$
$$
= \frac{C_k(s + \alpha_k) + E_k\beta_k}{(s + \alpha_k)^2 + \beta_k^2} \tag{1.40}
$$

where $\beta_k = \sqrt{\omega_k^2 - \alpha_k^2}$ and $E_k = \frac{D_k - C_k\alpha_k}{\beta_k}$. Hence,

$$
\mathcal{L}^{-1}\left\{ \frac{C_k s + D_k}{s^2 + 2\alpha_k s + \omega_k^2} \right\} = e^{-\alpha_k t}\left\{ C_k \cos \beta_k t + E_k \sin \beta_k t \right\}. \tag{1.41}
$$

Therefore, from (1.39) and (1.41), the complete step response of the closed-loop system can be written as

$$
y(t) = A + \sum_j B_j e^{-\sigma_j t} + \sum_k e^{-\alpha_k t}\left\{ C_k \cos \beta_k t + E_k \sin \beta_k t \right\}. \tag{1.42}
$$

Let us discuss each term of the above equation separately.

- The first term represents a step function and is always bounded
- The term $e^{-\sigma_j t}$: For this term to be bounded $\sigma_j \geq 0$. See also Fig. 1.13.
- Sine and cosine terms: These terms lead to an oscillating response and

  If $\alpha_k = 0$, oscillations are steady

  If $\alpha_k > 0$, oscillations will have a decreasing amplitude

  If $\alpha_k < 0$, oscillations will have an increasing amplitude.

  See also Fig. 1.14. For this term to be bounded, $\alpha_k \geq 0$.

  Note that just a single unbounded term can make the whole response unbounded. To this end, the condition for bounded output is that

$$
-\sigma_j, -\alpha_k \leq 0. \tag{1.43}
$$

In other words, the roots of the characteristic equation should not have positive real parts. That is, no poles of the system can be in the right half of the s-plane which is shown in Fig. 1.15.

An special case occurs when there are closed-loop poles repeating on the imaginary axis. A complex conjugate pair, $s = \pm j\omega_k$, occurs on the imaginary axis if there is a term $(s^2 + \omega_k^2)$ in the denominator of $M(s)$.

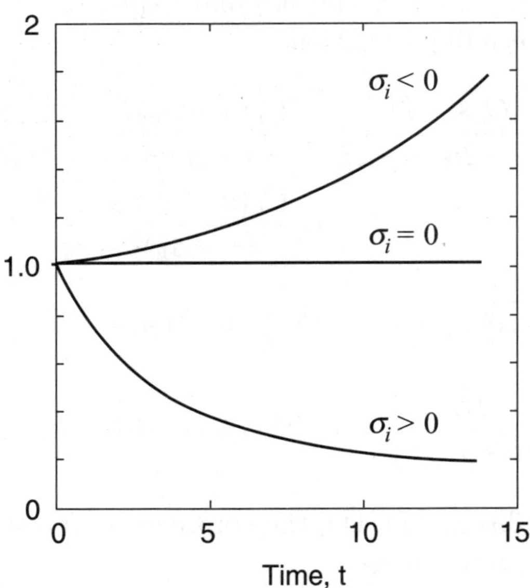

Figure 1.13: Purely exponential component.

Since $\alpha_k = 0$, according to (1.39), this will yield an steadily oscillating response and the system will be marginally stable. In contrast, if there is a $(s^2 + \omega_k^2)^2$ term in the denominator of $M(s)$, there will be multiple order poles at $s = \pm j\omega_k$. This will introduce an additional term of the following form making the system unstable. A term of the form

$$t \left\{ C_k \cos \beta_k t + E_k \sin \beta_k t \right\}$$

gives an oscillating response of which amplitude grows as time progresses.

Since they only exist in the form of complex conjugate pairs, if any complex roots repeat on the imaginary axis for any other reason than for multiple order poles at $s = \pm j\omega_k$, that will only be at the origin. An special case occurs when there are closed-loop poles repeating at the origin. If there is a root repeating at the origin, α_k and ω_k will both be zero in the last term of (1.39). Now,

$$\mathcal{L}^{-1} \left\{ \frac{C_k s + D_k}{s^2} \right\}$$

gives rise to following two terms in the system response

$$C_k + t D_k.$$

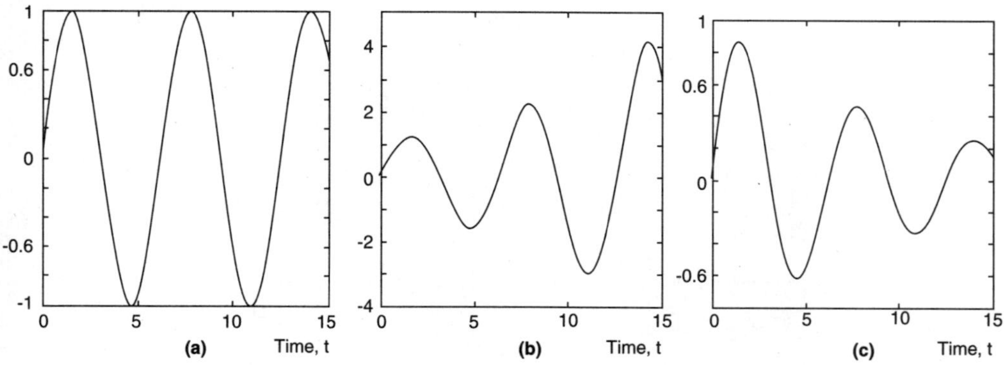

Figure 1.14: Oscillatory components: (a) $\alpha_k = 0$, (b) $\alpha_k < 0$, (c) $\alpha_k > 0$.

Since tD_k term grows indefinitely as $t \to \infty$, the total response will be unbounded. Hence, a more complete condition for bounded output is that the roots of the characteristic equation should not either have positive real parts or they can not repeat on the imaginary axis.

Example 1.1:

$$M(s) = \frac{30}{(s+1)(s+2)(s+3)} \qquad \text{Stable.}$$

$$M(s) = \frac{30(s+1)}{(s-2)(s^2+2s+3)} \qquad \text{Unstable due to the pole at } s = 2.$$

$$M(s) = \frac{30s}{(s+2)(s^2+4)} \qquad \text{Marginally stable due to poles at } s = \pm j2.$$

$$M(s) = \frac{30}{(s^2+4)^2(s+3)} \qquad \text{Unstable due to repeating poles at } s = \pm j2.$$

$$M(s) = \frac{30}{s^2(s+3)} \qquad \text{Unstable due to repeating pole at the origin.}$$

The stability characteristics of a linear time-invariant system can be determined from the system's characteristics equation. Routh's stability criterion that will be discussed shortly is a mathematical shortcut that provides a means for determining the stability without evaluating the roots.

1.8.1 Pole-zero Cancelation

Consider a system characterized by

$$\frac{Y(s)}{R(s)} = \frac{(s+a)G(s)}{(s+a)(1+G(s)H(s))}$$

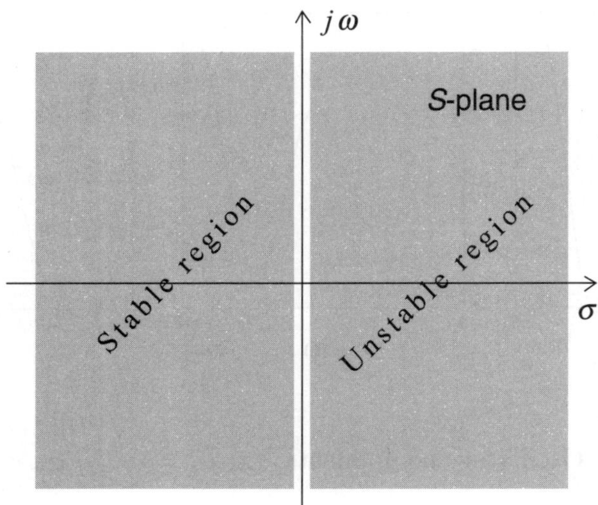

Figure 1.15: Stable and unstable regions in the s-plane.

where a is a constant. Here, we note a pole-zero cancelation that can be affected. Generally speaking about pole-zero cancelation, even though there can be situations where pole-zero cancelation can be affected we may not want to cancel a pole and a zero out of a transfer function. Caution is important. The poles and zeros may be close but not exact. If they are, for instance, on the right half of the s-pane, we may still have an unstable system even though the simplified transfer function after affecting pole-zero cancelation will not indicate it.

The concepts of controllability and observability will be discussed in chapter 3. If a system is either controllable and not observable or observable and not controllable, then pole-zero cancelation is possible in the closed-loop transfer function. However, due to aforementioned reasons, we may not want to cancel a pole and zero out of the transfer function. It is worth mentioning that most well-formed systems are both controllable and observable.

1.8.2 Routh's Stability Criterion

Let the characteristic equation of the closed-loop system be

$$1 + G(s)H(s) = a_0 s^n + a_1 s^{n-1} + a_2 s^{n-2} + \cdots + a_n. \qquad (1.44)$$

Observe the following tabular array that is also called Routh's array.

Table 1.1: Routh's array

s^n	a_0	a_2	a_4	a_6	$\cdots$
s^{n-1}	a_1	a_3	a_5	a_7	$\cdots$
s^{n-2}	A_1	A_2	A_3	$\ldots\ldots$	
s^{n-3}	B_1	B_2	B_3	$\ldots\ldots$	
$\vdots$	$\vdots$				
s^0	a_n				

where

$$A_1 = \frac{a_1 a_2 - a_0 a_3}{a_1}, \;\; A_2 = \frac{a_1 a_4 - a_0 a_5}{a_1}, \;\; \cdots \text{ and so on}$$

$$B_1 = \frac{A_1 a_3 - a_1 A_2}{A_1}, \; B_2 = \frac{A_1 a_5 - A_3 a_1}{A_1}, \cdots \text{ and so on}$$

Routh's stability criterion states that for a system to be stable, it is necessary and sufficient that each term of the first column of the Routh array be positive (assuming $a_0 > 0$). Furthermore, the number of sign changes in the 1st column corresponds to the number of roots of the characteristics equation in the right half of the s-plane. The proof is based on *continuous fraction test* and is omitted.

For example, consider the following characteristics equation

$$1 + G(s)H(s) = s^3 + 4s^2 + 8s + 12 = 0. \tag{1.45}$$

The Routh array for this system is shown in Table (1.2).   Since there are

Table 1.2: Routh's array for $1 + G(s)H(s) = s^3 + 4s^2 + 8s + 12$

s^3	1	8
s^2	4	12
s^1	5	0
s^0	12	0

no changes of sign in the 1st column, all the roots of the characteristics equation have negative real parts and hence the system is stable.

A special case arises when there is a zero coefficient in the 1st column. For instance, consider the characteristics equation

$$1 + G(s)H(s) = s^4 + s^3 + 2s^2 + 2s + 5 = 0. \qquad (1.46)$$

Routh's array is shown in Table (1.3).

Table 1.3: Routh's array for $1 + G(s)H(s) = s^4 + s^3 + 2s^2 + 2s + 5$

s^4	1	2	5
s^3	1	2	
s^2	0	5	
$\vdots$	?		

Zero in the 1st column makes it impossible to complete the array. To overcome this, one may modify the original characteristics equation, for instance, by multiplying it by (s+1). This introducing of an additional negative pole (pole on the left half of the s plane) will not alter the stability result and may make it possible to construct a complete array as follows. The modified characteristic equation,

$$(s + 1)\,[1 + G(s)H(s)] = s^5 + 2s^4 + 3s^3 + 4s^2 + 7s + 5 = 0 \qquad (1.47)$$

yields the Routh's array given in Table (1.4). There are two sign changes

Table 1.4: Routh's array for $1 + G(s)H(s) = s^5 + 2s^4 + 3s^3 + 4s^2 + 7s + 5$

s^5	1	3	7
s^4	2	4	5
s^3	1	4.5	
s^2	-5	5	
s^1	5.9		
s^0	5		

in the 1st column and hence two roots with positive real parts. This concludes that the original system is unstable.

One important application of Routh's criterion is to find the range of values of adjustable control gains for which the system is stable. Once such a range is found, the designer can fine tune system performance by iterating on the adjustable gains.

Example 1.2: In the following unity feedback system, K is an adjustable parameter. Determine the range of K for which the system is stable.

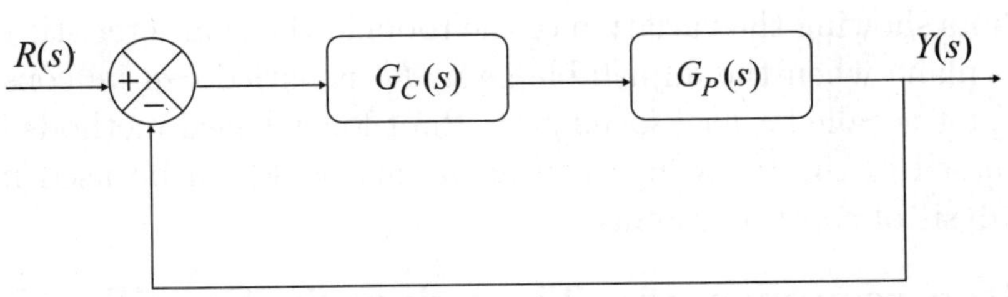

Figure 1.16: System for example 1.3.

The closed-loop transfer function is

$$M(s) = \frac{C(s)}{R(s)} = \frac{K(s+2)}{s(s+5)(s^2+2s+5)+K(s+2)} \tag{1.48}$$

The characteristics equation is

$$s^4 + 7s^3 + 15s^2 + (25+K)s + 2K = 0. \tag{1.49}$$

The Routh's array is given in Table (1.5).

Table 1.5: Routh's array for Example 1.2

s^4	1	15	$2K$
s^3	7	$25+K$	
s^2	$\frac{80-K}{7}$	$2K$	
s^1	$\frac{(80-K)(25+K)-98K}{80-K}$		
s^0	$2K$		

For stability

$$80 - K > 0$$
$$(80 - K)(25 + K) - 98K > 0$$
$$\text{and } K > 0$$

yield $0 < K < 28.1$ to be the range for which the system is stable. Note from the closed-loop transfer function that there will be zero response for $K = 0$. For $K = 28.1$, the system step response will be steadily oscillating. The closed-loop system in this situation, is also said to be marginally stable. Some authors define this situation also to be stable in that the system response is bounded and does not grow indefinitely.

Plots showing the variation of the roots of the characteristic equation in the s-plane when the adjustable gain, K, is varied can be constructed. Such a plot is called a *root locus plot*. Root locus-based methods that will be discussed in the following chapter of this book can be used in design and analysis of control systems.

1.9 Step-response and Time-domain Specifications

Control systems are designed to meet the design specifications. Design specifications specify the expected transient and steady-state behaviors of the system, i.e., the ability of the system to maintain the error negligible as input changes, steady-state accuracy after the transients have died out, and system stability. In general, the transient response is based on the the denominator and the steady-state response on the numerator of the closed-loop transfer function.

Consider a prototype second-order system with unity feedback with the open-loop transfer function being

$$G(s) = \frac{Y(s)}{E(s)} = \frac{\omega_n^2}{s(s + 2\zeta\omega_n)}$$

where ω_n is the undamped natural frequency of the system and ζ is the damping ratio. The closed-loop transfer function of the system is

$$\frac{Y(s)}{R(s)} = \frac{\omega_n^2}{s^2 + 2\zeta\omega_n s + \omega_n^2}. \tag{1.50}$$

Many systems can be expressed in this form or can be approximated by second-order systems. Some higher-order systems are composed of cascaded combinations of second-order systems.

The response of a control system when the input is a unit-step function is called unit-step response. For a unit-step input, $R(s) = 1/s$, the Laplace transform of the output is

$$Y(s) = \frac{\omega_n^2}{s(s^2 + 2\zeta\omega_n s + \omega_n^2)}.$$

Taking inverse Laplace transform that is omitted for brevity yields

$$y(t) = 1 - \frac{e^{-\zeta\omega_n t}}{\sqrt{1-\zeta^2}} \sin(\omega_n\sqrt{1-\zeta^2}t + \cos^{-1}\zeta) \quad t \geq 0. \tag{1.51}$$

1.9.1 Undamped Natural Frequency and the Damping Ratio

The characteristic equation of the second order system (1.50) is given by

$$s^2 + 2\zeta\omega_n s + \omega_n^2 = 0.$$

The two roots are

$$s_1, s_2 = -\zeta\omega_n \pm j\omega_n\sqrt{1-\zeta^2}.$$

As seen in (1.51), $\zeta\omega_n$ is the term that is multiplied to t in the exponential term of the unit-step response. Hence, this term that is known as the *damping factor* controls the rate of rise or decay. When the closed-loop system is critically damped, i.e., $\zeta = 1$ and the roots are real and equal, damping factor is simply ω_n. Hence, ζ can be regarded as the damping ratio and

$$\zeta = \text{damping ratio} = \frac{\text{damping factor}}{\text{damping factor at critical damping}}.$$

When $\zeta = 0$, the system is undamped and the roots of the characteristic equation are purely imaginary. According to (1.51), the unit-step response is purely sinusoidal with the frequency of ω_n. To this end, ω_n is termed the undamped natural frequency.

1.9.2 Transient Response Specifications

When $\zeta = 1$, the response is the fastest without any overshoot. Then, the response is called *critically damped*. When $\zeta > 1$, the response is *overdamped* and non-oscillatory. Such a response may often considered to be too sluggish. When $\zeta < 1$, the response is oscillatory and is said to be *underdamped*. Fig. 1.17 shows the unit-step response, (1.51), for various values of ζ when ω_n is held constant.

We can now define the following transient response parameters.

- *Rise Time, t_r*: is the time required for the unit step response to rise from 10% to 90% or 5% to 95% or 0% to 100% depending on the situation. For underdamped 2nd order systems, 0-100% is commonly used.
- *Peak Time, t_p*: is the time required for the response to reach the first peak of the overshoot.

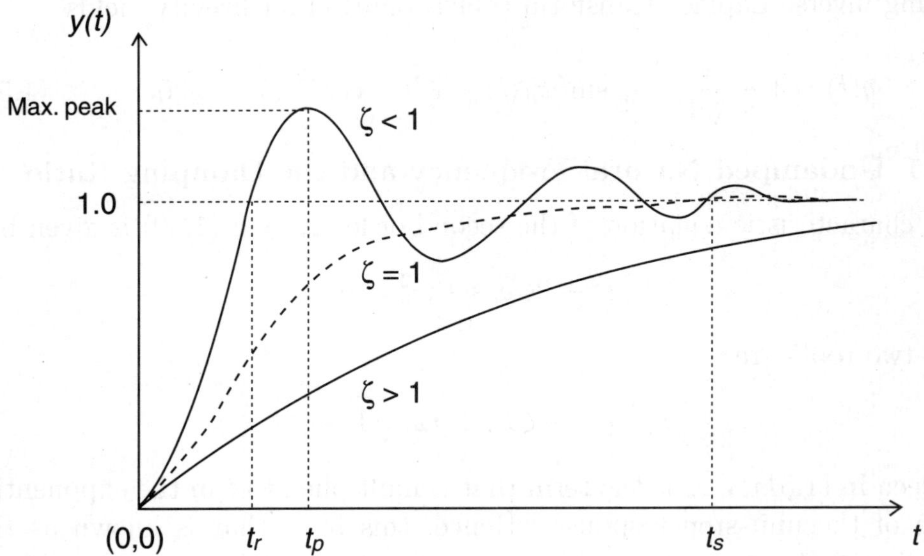

Figure 1.17: Transients of the unit step response of a second-order system.

- *Maximum Overshoot, M_p:* is the maximum peak value of the response measured from unity.
- *Settling Time, t_s:* is the time required for the response to reach and stay within a range about the final value of a size specified as an absolute percentage of the final value. For a second-order system, the response is usually within 2% after t_s.

The above definitions of transient response parameters are general. For the prototype second-order system, by analyzing (1.51), following results can be obtained.

$$t_r = \frac{0.8 + 2.5\zeta}{\omega_n}$$

$$t_p = \frac{\pi}{\omega_n \sqrt{1 - \zeta^2}}$$

$$M_p = \exp\left\{\frac{-\pi\zeta}{\sqrt{1 - \zeta^2}}\right\}$$

$$t_s \approx \frac{4}{\zeta\omega_n}.$$

Time domain specifications are quite important since the control systems must exhibit acceptable time responses. This means that the control system being designed must be tuned until the transient response is satis-factory.

1.9.3 Steady State Error

Recall the standard error-feedback closed-loop control system (Fig. 1.6) where $r(t)$ is the reference input and $y(t)$ is the output. The error of the system may be defined as

$$e(t) = \text{reference signal} - \text{feedback signal}. \tag{1.52}$$

And hence

$$E(s) = R(s) - H(s)Y(s). \tag{1.53}$$

The steady-state error is defined as

$$e_{ss} = \lim_{t \to \infty} e(t). \tag{1.54}$$

Using the final value theorem

$$e_{ss} = \lim_{t \to \infty} e(t) = \lim_{s \to 0} sE(s). \tag{1.55}$$

For a unity feedback, i.e., $H(s) = 1$, system

$$E(s) = R(s) - G(s)E(s) \quad \text{and} \quad E(s) = \frac{R(s)}{1 + G(s)}.$$

Hence, the preceding equation becomes

$$e_{ss} = \lim_{s \to 0} sE(s) = \lim_{s \to 0} \frac{sR(s)}{1 + G(s)}. \tag{1.56}$$

Clearly, e_{ss} depends on the characteristics of the forward-path transfer function, $G(s)$.

For a discrete system, the equation becomes

$$e_{ss} = \lim_{k \to \infty} e(k) = \lim_{z \to 1}(1 - z^{-1})E(z) = \lim_{z \to 1}(1 - z^{-1})\frac{R(z)}{1 + G(z)}. \tag{1.57}$$

Note that the final value theorem for z-transform was used in the preceding equation.

Chapter 2

Graphical Methods in Control Theory

2.1 Introduction

Several established graphical methods are available for control system design and analysis. If the system model is available, one can solve these differential equations to obtain the time response of the system. However, if the response does not meet the design specifications, it is not easy to figure out, from this solution, what physical parameters in the system should be changed (and how) to improve the response.

Besides, the designer can check system stability by solving the characteristics equation or using Routh's stability criterion without solving the characteristics equation. Yet, this does not indicate the degree of stability, i.e., the amount of overshoot, settling time etc. of the controlled variables. Why is ensuring smaller overshoot important? Consider a motor control problem. Large overshoot may even drive the motor into the saturation region inviting undesirable nonlinearities that may even cause divergence. Chattering caused by poor transients can cause poor system performance and reduce lifetime. In a passenger vehicle, large overshoots and longer settling time may give rise to poor ride comfort levels. Shorter settling time and rise times are generally preferred for a faster response to ensure better controller performance.

The graphical methods such as root locus method, Bode plots, Nyquist plots, etc. that will be discussed in this chapter not only indicate whether the system is stable or not but also show the degree-of-stability. A number of commercially available CAD packages (eg: MATLAB, ICECAP-PC or TOTAL-PC) can also be used if the mathematical model of the system is available.

2.1.1 Open-loop Transfer Function

If the feedback loop of an error feedback control system is made open at the comparator, the system overall transfer function becomes

$$G(s)H(s) \qquad (2.1)$$

where $G(s) = G_P(s)G_C(s)$ using the familiar notation. (2.1) is called the open-loop transfer function. Since there is no feedback signal to be compared, the comparator becomes a virtual short circuit. It is noteworthy that the poles of the closed-loop transfer function or the roots of $1 + G(s)H(s) = 0$ are related to the zeros and poles of the open-loop transfer function, $G(s)H(s)$.

This important observation is often exploited in the graphical methods discussed in the sequel and the closed-loop systems will be analyzed by the use of their open-loop transfer functions and the characteristics.

2.2 Root Locus Method

The key advantage of the root locus method is that the actual time response can be obtained using inverse Laplace transform when the roots of the characteristics equation are known. The root locus is a plot of the roots of the characteristics equation of the closed-loop system as a function of the gain, K, of the open-loop transfer function (OLTF). The gain K, which is also called the loop sensitivity is adjustable to obtain the desired performance specifications and is often attached to the controller block of the control system. The important advantage of the root locus is that it shows how the roots of the characteristics equation changes with the adjustable gain, K, on the s-plane giving a clear indication of the effect of gain adjustment.

The roots of the characteristics equation of the closed-loop system are related to the zeros and poles of the OLTF and to the loop sensitivity. This is explained as follows. With K being the loop sensitivity that is usually adjustable, the OLTF takes the form

$$G(s)H(s) = \frac{K(s + a_1)(s + a_2) \cdots (s + a_w)}{(s + b_1)(s + b_2) \cdots (s + b_u)} = K\frac{N(s)}{D(s)}. \qquad (2.2)$$

Therefore, the closed-loop characteristics equation takes the following form

$$1 + K\frac{N(s)}{D(s)} = 0. \qquad (2.3)$$

When the loop sensitivity, K, goes from zero to infinity (K: $0 \rightarrow \infty$), roots of the characteristics equation moves from open-loop poles to open-loop zeros (i.e., closed-loop poles: open-loop poles $\rightarrow$ open-loop zeros). One main advantage of the root locus method is that we need only the positions of the zeros and poles of the OLTF to obtain the locus of closed loop poles.

2.2.1 Construction of the Root Locus

Two options are available to construct a root locus, namely, manually and using CAD software. To take a look at manual construction of a root locus, consider the analog implementation of a position control system shown in Fig. 2.1 where A is the adjustable gain of an amplifier, J and B, respectively, are the inertia and damping.

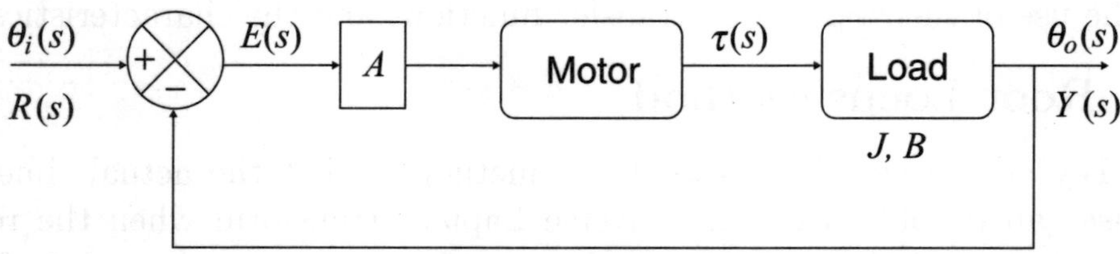

Figure 2.1: An analog position control system.

Assume that the transfer function of the plant (motor plus the mechanical load) has been given as

$$G_P(s) = \frac{1}{Js(s + B/J)}. \tag{2.4}$$

According to Fig. 2.1, $G_C(s) = A$. For an unity feedback system, $H(s) = 1$. Therefore, the OLTF is

$$G(s)H(s) = G(s) = \frac{A/J}{s(s + B/J)} = \frac{K}{s(s + b)} \tag{2.5}$$

where $K = A/J$ is adjustable by adjusting A and $b = B/J$.

Assume $b = 2$. Now the closed-loop transfer function becomes

$$\frac{C(s)}{R(s)} = \frac{G(s)}{1 + G(s)} = \frac{K}{s(s + 2) + K}. \tag{2.6}$$

Roots of the characteristics equation, $1 + G(s) = 0$, are shown in Table 2.1 for a number of values of K. Now, the variation of the two roots,

Table 2.1: Roots of $s(s+2) + K = 0$

K	$s_1 = -1 + \sqrt{1-K}$	$s_2 = -1 - \sqrt{1-K}$
0.0	0+j0	-2.0 -j0
0.5	-0.29+j0	-1.71-j0
$\vdots$	$\vdots$	$\vdots$
1.0	-1.0+j0	-1.0-j0
2.0	-1.0+j1.0	-1.0-j1.0
$\vdots$	$\vdots$	$\vdots$
50.0	-1.0+j7.0	-1.0 -j7.0

s_1 and s_2, can be plotted in the s-plane for different values of K as in Fig. 2.2 for $0 < K < \infty$. For negative values of K, there is always a root in the right-half of the s-plane. Since the topic is relative or degree of stability, the range of K for which the system is unstable is normally omitted in these plots. These plots are defined as the *root locus* plots. Once this plot is obtained, the roots that best fit the system performance can be selected with the corresponding value of K. Therefore, the root locus can be used to determine the variation in system performance with respect to a variation in sensitivity, K.

With some theoretical background (Recall equation (1.42) from chapter 1, the generalized step response of a system whose transfer function is $M(s) = \frac{C(s)}{R(s)}$) and experience, one may be able to predict the variation in system time response with respect to a variation in loop sensitivity, K. This is relatively much easier for lower-order systems such as the underlining motor position control problem, which is a second-order system. In this example, frequency of the transient response is zero for $K \leq 1$ as there are no complex roots and hence the trigonometric terms in the step response. For $K > 1$, one can expect an increase in frequency of the transient response as the imaginary component of the complex roots increases with K. This second-order system will also show increasing overshoot of the time response for increasing values of K. To that end, the root locus of a control system can be analyzed to determine the variation of its time response in response to a variation of its loop sensitivity.

Manual construction of the root locus of higher-order systems is rather thorny. As CAD software packages are readily available now-a-days,

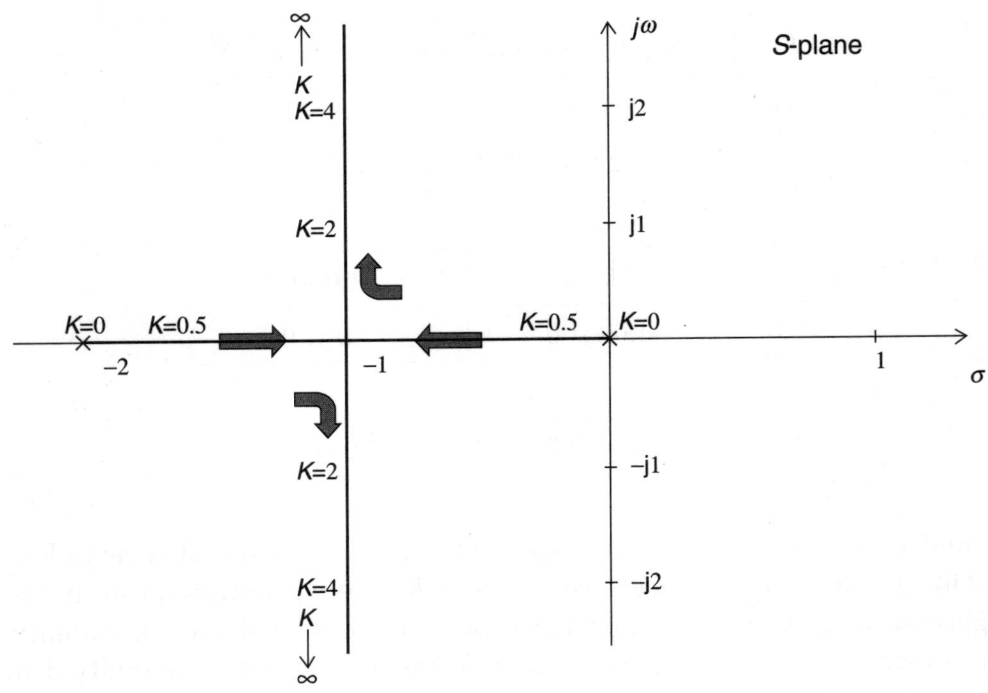

Figure 2.2: Variation of the roots with K in the s-plane.

discussion on manual construction will be given a minimum priority in this book. To that end, the following rules that can be used in constructing a root locus manually are mentioned without proof for brevity.

- **Rule 1:** *Symmetry*: As the roots are real or complex conjugate pairs, root locus is symmetrical about the real axis.

- **Rule 2:** *Number of loci*: Let the number of OLTF poles and OLTF zeros be n and m, respectively. In general, $n > m$. The number of loci will be n as each loci will start from an OLTF pole. There will be $(n - m)$ loci ending at infinity.

- **Rule 3:** *Real axis loci*: Some of the loci will lie on the real axis. A point on the real axis will lie on the root locus if and only if the sum of the OLTF poles and zeros to the right of the point is odd.

- **Rule 4:** *Angle of asymptotes*: The number of asymptotes $= n - m$. The angle of asymptotes is given by

$$\theta = \frac{2q + 1}{n - m} \times 180^0 \quad \text{for } K > 0$$

$$\theta = \frac{2q}{n-m} \times 180^0 \quad \text{for } K < 0$$

where $q = 0, 1, 2, \cdots, n - m - 1$.

- **Rule 5:** *Center of asymptotes*: The point where the asymptotes touch the real axis is given by

$$P_c = \frac{\sum \text{real parts of poles} - \sum \text{real parts of zeros}}{n-m}. \qquad (2.7)$$

Note that in constructing root-locus, the OLTF is considered. Therefore, the preceding equation refers to open-loop poles and zeros.

- **Rule 6:** *Break away/break-in points*: Break-away (or break-in) points are the points at which the root locus leaves (enters) the real axis.
For instance, if there are two adjacent poles on the real axis and the real axis is a part of the root locus, one minimum break-away point exists in between adjacently placed poles. Also, if there are two adjacently placed zeros on the real axis and the real axis is a part of the root locus, one minimum break-in point exists in between these zeros.
Following steps can be followed to determine the break-away points. Write the characteristics equation $1 + G(s)H(s) = 1 + KG_1(s)H_1(s) = 0$ where G_1 and H_1 are suitably defined in a way K is highlighted in the equation. This Allows one to write $K = -1/G_1(s)H_1(s)$. The roots of the equation $dK(s)/ds = 0$ are the break-away points. If $K > 0$ for any root of $dK(s)/ds = 0$, that root can be considered as a valid break-away point.
Eg: Consider $1 + G(s)H(s) = s^3 + 5s^2 + 6s + K = 0$. This yields

$$K = -s^3 - 5s^2 - 6s \qquad (2.8)$$

and

$$\frac{dK}{ds} = -3s^2 - 10s - 6. \qquad (2.9)$$

The roots of $dK/ds = 0$ are

$$s_1, s_2 = \frac{-10 \pm \sqrt{(10^2 - 4 \times 3 \times 6)}}{2 \times 3}$$
$$= -0.79, -2.55.$$

Now, as $K(s_1) = K(-0.79) = (-0.79)^3 - 5(-0.79)^2 - 6(-0.79) > 0$ and $K(s_2) = K(-2.55) < 0$, $s = -0.79$ is a valid break-away point.

- **Rule 7:** *Intersection of root locus with the imaginary axis*: Use Routh's criterion to find the value of K for which the system is marginally stable. This is called critical value of K for stability. Now, solve the closed-loop characteristics equation to find the corresponding purely imaginary roots that are the points of intersection of the root locus with the imaginary axis.

  A shortcut is to solve the auxiliary equation.

  Eg: Routh array for $s^3 + 2s^2 + 2s + K = 0$ is shown below.

$$
\begin{array}{c|cc}
s^3 & 1 & 2 \\[4pt]
s^2 & 2 & K \\[4pt]
s^1 & \frac{4-K}{2} & 0 \\[4pt]
s^0 & K &
\end{array}
$$

According to the Routh array, the critical value of K for stability is $K = 4$. The crossover points on the imaginary axis can be obtained by solving the auxiliary equation as follows

$$2s^2 + K = 0$$

$$s^2 = -\frac{K}{2} \quad \Rightarrow s = \pm j\sqrt{2}.$$

- **Rule 8:** *Angle of departure/arrival*: The angle of departure of the root locus from a complex pole is given by

$$\theta_d = 180^0 + \arg[G(s)H(s)] \tag{2.10}$$

  where $\arg[G(s)H(s)]$ is the angle of $G(s)H(s)$ excluding the pole where the angle is calculated. The angle of arrival at a complex zero is

$$\theta_a = 180^0 - \arg[G(s)H(s)] \tag{2.11}$$

  where $\arg[G(s)H(s)]$ is the angle of $G(s)H(s)$ excluding the zero where the angle is calculated.

Construction of the root locus using computer-aided design (CAD) software is highly recommended for higher-order systems. For instance, Matlab is widely used in industry and academia alike now-a-days.

Example 2.1: Consider the characteristics equation $s^3 + 2s^2 + 2s + K = 0$. The equivalent $G(s)H(s)$ is obtained by dividing both sides by the terms that do not contain K. We have

$$G(s)H(s) = \frac{K}{s^3 + 2s^2 + 2s}. \tag{2.12}$$

Based on $G(s)H(s)$, the root locus of $s^3 + 2s^2 + 2s + K = 0$ will be plotted
first by the use of the rules for manual construction and then using Matlab
software package.

- Step-1: From (2.12), it follows that $n = 3$ and $m = 0$. And, hence

$$\text{Number of loci} = n - m = 3. \tag{2.13}$$

- Step-2: The poles of the OLTF in the s-plane will be at $(0, j0)$, $(-1, -j1)$,
  and $(-1, j1)$. Since there are no zeros of the OLTF, the three loci that
  start at the open-loop poles will end up at infinity.
- Step-3: Real axis locus exists for

$$-1 < \sigma < 0$$
$$-\infty < \sigma < -1.$$

- Step-4:

$$n - m = 3 \quad \text{and} \quad q = 0, 1, 2$$
$$\theta_1 = \frac{2 \times 0 + 1}{3} \times 180^0 = 60^0, \qquad \theta_2 = \frac{2 \times 1 + 1}{3} \times 180^0 = 180^0$$
$$\theta_3 = \frac{2 \times 2 + 1}{3} \times 180^0 = 300^0.$$

- Step-5: The point where the asymptotes touch the real axis is given
  by

$$P_c = \frac{(0 - 1 - 1) - 0}{3} = -0.667. \tag{2.14}$$

See Fig. 2.3.
- Step-6: Calculating breakaway points is not required in this example.
- Step-7: Applying Routh's criterion to $s^3 + 2s^2 + 2s + K = 0$, and
  by solving the auxiliary equation, we have the critical value of K for
  stability at $K = 4$, and the corresponding crossover points on the
  imaginary axis of the s-plane are at $\pm j\sqrt{2}$.
- Step-8:

$$G(s)H(s)|_{s=-1+j} = \frac{K}{(-1+j)(-1+j+1+j)}$$
$$= \frac{K}{\sqrt{2}}\angle -225^0.$$

Therefore, the angle of departure from $s = -1 + j$ is

$$\theta_d = 180^0 + \arg[G(s)H(s)] = 180^0 - 225^0 = -45^0.$$

By symmetry, the angle of departure from $s = -1 - j$ is $+45^0$.

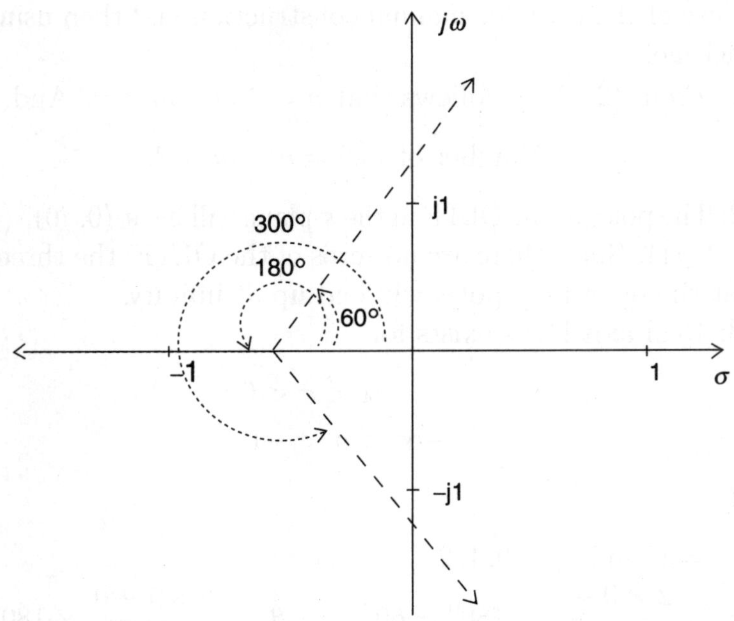

Figure 2.3: The asymptotes.

The complete root locus is given in Fig. 2.4.

MATLAB command to construct the underlining root locus is

$$\mathrm{sys} = \mathrm{tf}([1], [1\ 2\ 2\ 0])$$

$$\mathrm{rlocus(sys).}$$

rlocus(sys) function calculates and plots the root locus of the system represented by the open-loop model, sys. Matlab adaptively selects a set of positive gains K and produces the plot of the root locus of the closed-loop transfer function shown in Fig. 2.5.

Alternatively, rlocus(sys, K) uses the user-specified vector K of gains to plot the root locus. The function, rlocfind returns the gain associated with a particular set of poles on the root locus.

$$[K, \mathrm{poles}] = \mathrm{rlocfind(sys).}$$

In the Matlab rlocus plot, now we can position the cross-hairs at any point and get the closed-loop pole values and the values of K at that point. For instance, at the point where the root locus crosses the imaginary axis, we get $K = 3.9538$ and poles at -0.0039±j1.4087 and -1.9923 that are very close to those calculated by the Routh's criterion.

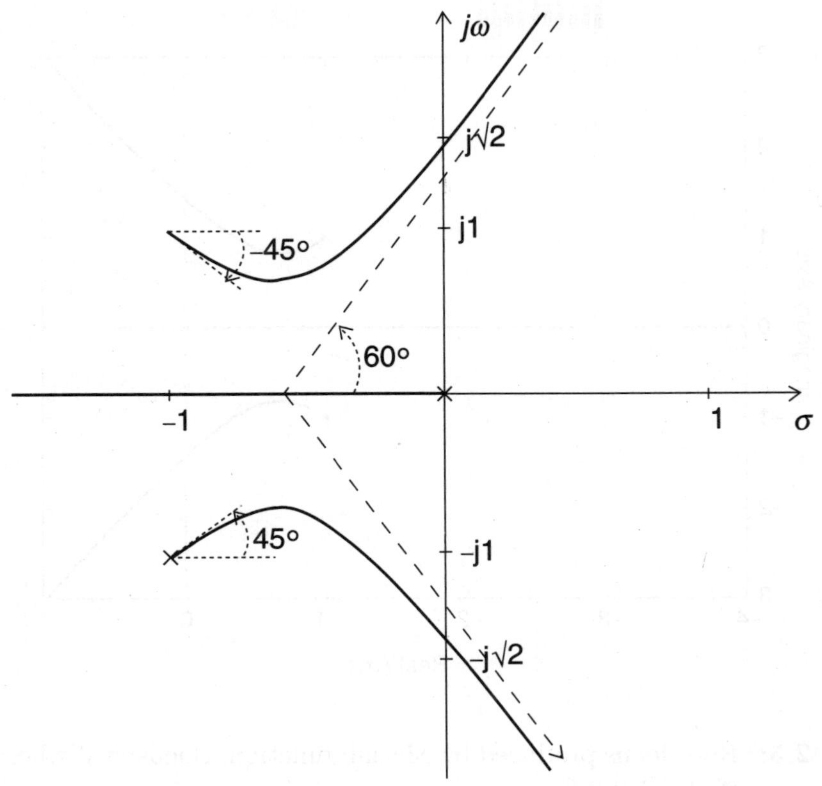

Figure 2.4: The root locus.

2.2.2 Reshaping the Root Locus–effects of Addition of Poles

If the root locus method-based design procedure, closed-loop pole placement by gain adjustment alone does not yield the desired performance,
one can consider reshaping the root locus by adding poles and zeros to the
OLTF. Effects of addition of poles to $G(s)H(s)$ are

- There is a change in shape of the root locus and it shifts towards the
  imaginary axis
- The intercept on the imaginary axis occurs for a lower value of K
  because asymptote's angle gets smaller (see *Rule 4*). This decreases
  the range of K.
- System becomes oscillatory (degree of stability will be lower)
- Settling time increases
- Therefore, a sluggish response can be changed to a quicker response
  by carefully introducing a pole

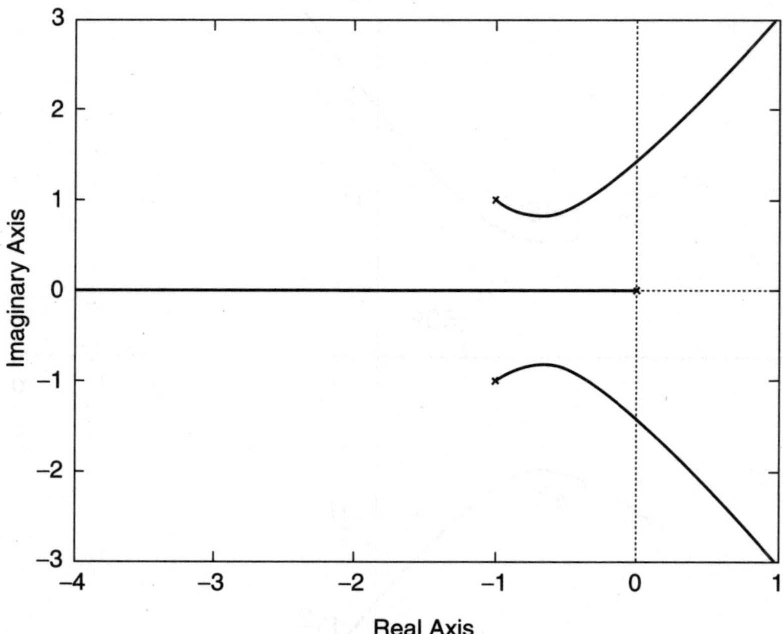

Figure 2.5: Root locus produced by Matlab function, rlocus(sys) where sys = tf([1], [1 2 2 0]).

2.2.3 Reshaping the Root Locus–effects of Addition of Zeros

- There is a change in shape of the root locus and it shifts away from the imaginary axis
- The range of K and the stability of the system is enhanced
- Settling time decreases

Eg: Consider $G(s)H(s) = \frac{K}{s(s+a)}$ with $a > 0$. The root locus of $1 + G(s)H(s) = 0$ is shown in Fig. 2.6.

Now, introduce a pole at $s = -b$ with $b > 0$. Hence

$$G(s)H(s) = \frac{K}{s(s+a)(s+b)}$$

and the root locus is given in Fig. 2.7.

The complex part of the root locus has bent towards the right-half of the s-plane. The angles of asymptotes have changed from 90^0 to 60^0. The intersect of the asymptotes on the real axis has also moved from $-a/2$ towards the imaginary axis. System becomes unstable when K exceeds a

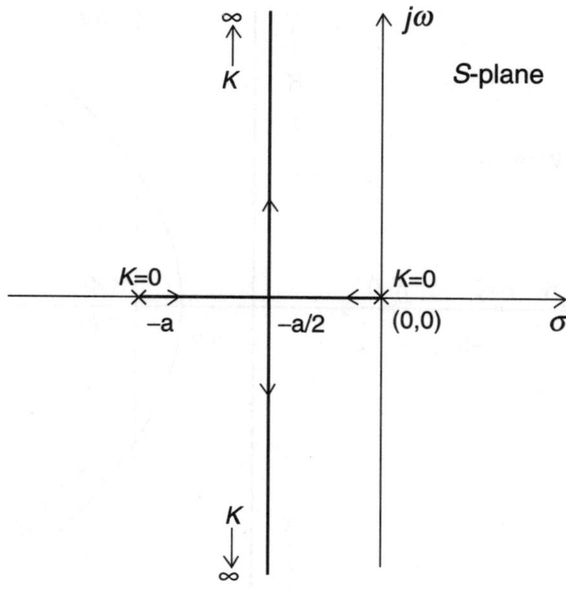

Figure 2.6: Root locus for $G(s)H(s) = K/s(s+a)$ where $a > 0$.

certain value whereas the system before introduction of the pole was always stable for $K > 0$. The stability of the 3rd order system is more acute than that of the 2nd order system.

Fig. 2.8 is the root locus of the system when the original system when a zero is added at $s = -b$ ($b > 0$). The complex part of the root locus is bent towards the left and form a circle. Thus, the relative stability is improved in the system by the addition of the zero.

2.2.4 Design Procedure Outline

- Step 1: Derive the open-loop TF (OLTF), $G(s)H(s)$, of the system and factor the numerator and the denominator.
- Step 2: Plot the zeros and poles of $G(s)H(s)$, in the s-plane and use the geometrical shortcuts (*Rule-1* through *Rule-8*) to determine the root locus. Or, use a CAD program as a simple command can draw the root locus when the OLTF is known.
- Step 3: Calibrate the locus in terms of the loop sensitivity, K. Now, when K is known, the exact roots of the closed-loop TF are known and vise versa.
- Step 4: When the roots are known, the system time response can be obtained using inverse Laplace transform, either manually or using a CAD program.

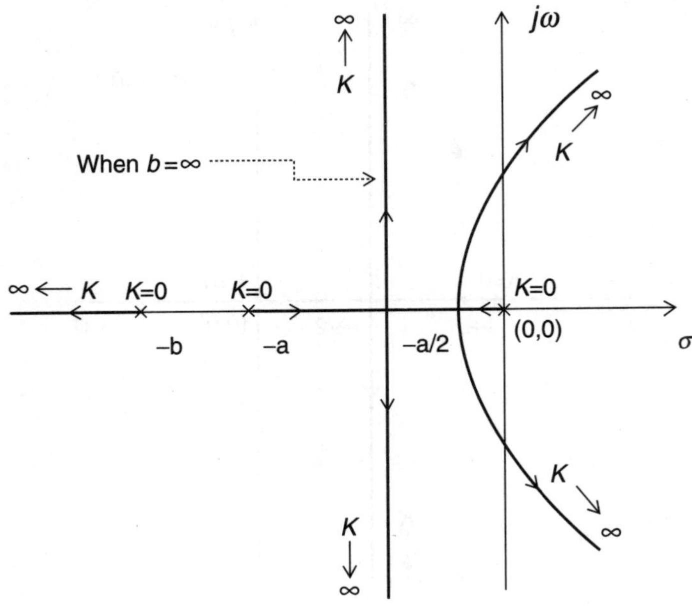

Figure 2.7: Root locus for $G(s)H(s) = K/s(s+a)(s+b)$ where $a,\ b > 0$.

- Step 5: If the response does not meet the performance specifications, guess the shape that the root locus must have to meet the specifications (reshaping of the root locus)
- Step 6: If the gain adjustment and reshaping alone des not yield the desired performance, insert a cascade compensator (eg. PID) into the system.

2.3 PID Controller (PID Compensator)

In an error feedback control system shown in Fig. 2.9, the controller transfer function defines the relationship between the error signal, $E(s)$, that is the input and the control command, $U(s)$, which is the output or the control input to the plant. Control law or the control strategy, in this case is expressed in the s-domain as $G_C(s)$, determines how the control commands are produced in a way the error is always regulated at zero (ideally speaking) while also ensuring a quality transient performance. Proportional controller that produces a control command that is directly proportional to the input error signal is one of the simplest control laws that is in use. The proportional (P) controller can be expressed either in the Laplace domain or in the time domain as

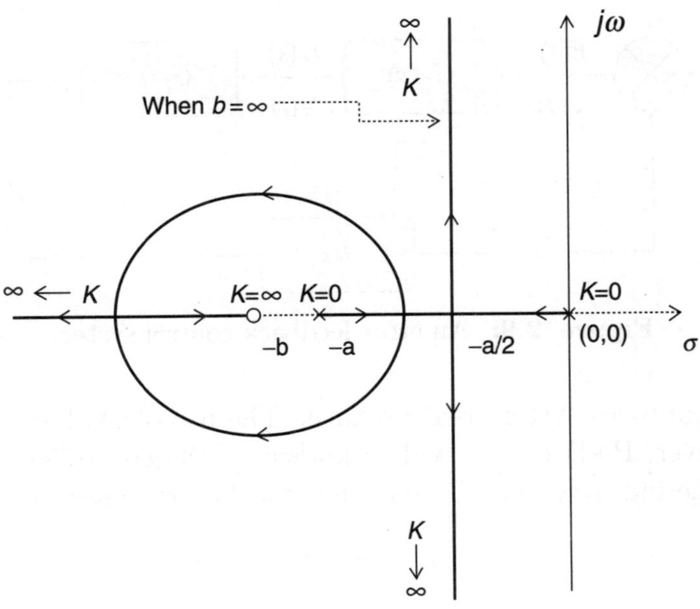

Figure 2.8: Root locus for $G(s)H(s) = K(s+b)/s(s+a)$ where $a,\ b > 0$.

$$G_C(s) = k_P \tag{2.15}$$

$$\text{or} \quad u(t) = k_P e(t) \tag{2.16}$$

where k_P is the proportional gain that is to be chosen by the designer. Consider a step change to the reference input at $t = t_0$ giving rise to a large error, $e(t)$, at t_0. With proportional control, the controller can introduce a large gain and hence decrease the rise-time. As a result, an speedy response can be expected initially as the output will follow the input faster. But the possible problem with a P-controller is increased maximum overshoot. Overshoot can invite undesirable nonlinearities such as saturation in motor control. As linear control theories may not be valid in these regions, this may even lead to instability sometimes.

A derivative (D) control action can be added to the control law to overcome the problem of large maximum overshoot yet ensuring a satisfactorily small rise time. Note that in P-control, the controller is only sensitive to the instantaneous value of error. The idea of D-control is to make the controller also respondent to the transients by introducing a term that is sensitive to the rate of change of error or the error rate. So, a PD type

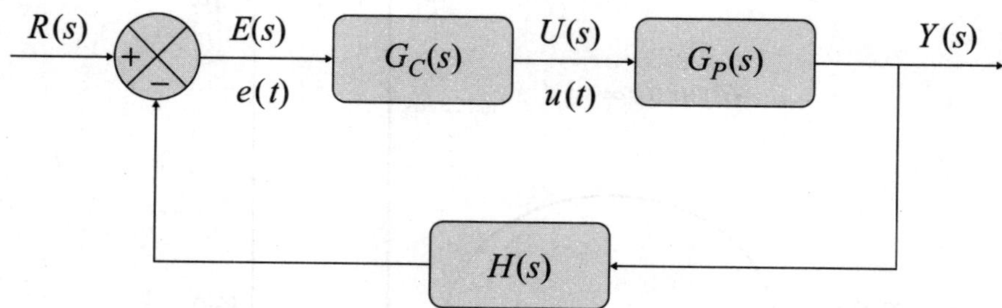

Figure 2.9: An error feedback control system.

controller improves system performance. The use of D-alone controller is rare. However, P+D that is widely known as PD controller is a popular option in the industry. The D-controller can be expressed as

$$G_C(s) = k_D s \tag{2.17}$$

$$\text{or} \qquad u(t) = k_D \frac{de(t)}{dt} = k_D \dot{e}(t) \tag{2.18}$$

where k_D is the derivative gain. After the transients have died out there may still remain a small yet considerable steady state error. This is also called the off-set. In the steady state where $\dot{e} = 0$, the D-control term is inactive. Furthermore, the P-control action may also be ineffective as k_P can be too small to effectively deal with small steady-state error. Therefore, one major problem of PD controller may be a considerably large steady-state error.

To reduce the steady-state error, one can use an integral (I) controller given below.

$$G_C(s) = k_I \frac{1}{s} \tag{2.19}$$

$$\text{or} \qquad u(t) = k_I \int_0^t e(\tau)d\tau. \tag{2.20}$$

Although an integral control is occasionally used by itself, a more common scheme is P+I (PI controllers). More generally, and in many control system applications, such as servo motor control systems, the combination of proportional, integral, and derivative control are used in the famous PID controller. PID controller has the transfer function

$$G_C(s) = k_P + k_D s + k_I \frac{1}{s} \tag{2.21}$$

where k_P, k_D, and k_I are the gains of the controller. These gains should be suitably chosen by the designer to achieve the desired control performance or the design specifications. Tuning the gains to the best values is sometimes achieved using trail and error. Empirical methods like Zeigler-Nichols method that will be discussed shortly are also used.

2.4 Implementation of the PID Controller

PID controller and its variants, namely, P, PD, and PI, can be implemented using analog circuits constructed by the use of summing, amplifying, integrating, differentiating circuits implemented using operational amplifiers, resisters, and capacitors. An op-amp circuit realization of the PI controller is given in Fig. 2.10. Note that the tail end is an inverting circuit.

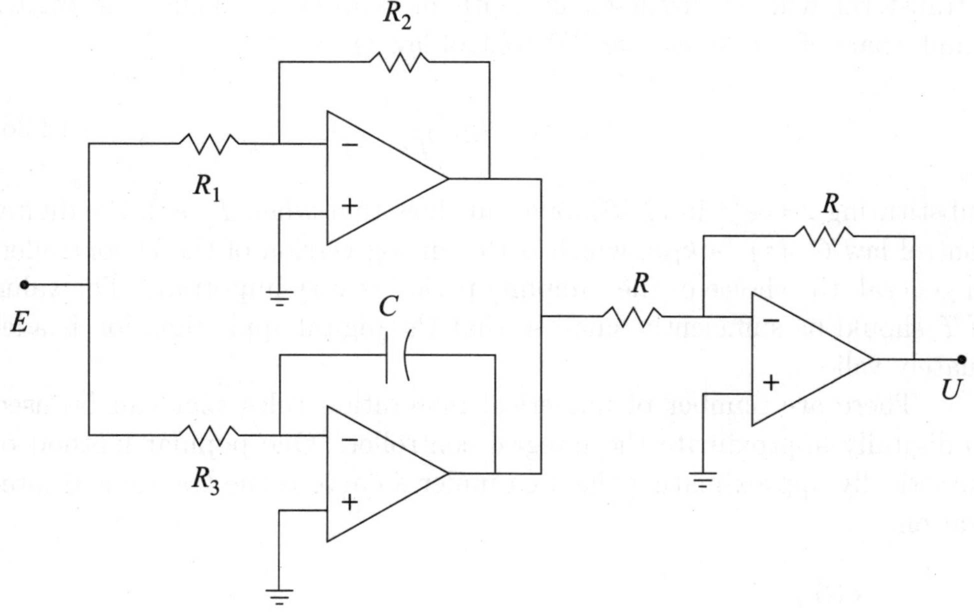

Figure 2.10: An Operational amplifier circuit realization of the PI controller.

The transfer function of this circuit is

$$G_C(s) = \frac{U(s)}{E(s)} = \frac{R_2}{R_1} + \frac{1}{R_3 C}\frac{1}{s}. \tag{2.22}$$

Thus the PI controller gains in terms of the circuit parameters are

$$k_P = \frac{R_2}{R_1} \qquad k_I = \frac{1}{R_3 C}. \tag{2.23}$$

In digital implementation of the PID controller, the time-derivative of error, $\dot{e}(t)$, at the time instant $t = kT$ where k is the time index and T is the sampling period may be approximated by backward-difference rule using the values of $e(t)$ measured at $t = kT$ and $(k-1)T$ as follows

$$\dot{e}(t)\,|_{t=kT} = \frac{e(kT) - e[(k-1)T]}{T}. \tag{2.24}$$

To find the z-transfer function of the derivative operator, let us take the z-transform of both sides of the preceding equation as follows

$$\mathcal{Z}\left\{\dot{e}(t)\,|_{t=kT}\right\} = \frac{1}{T}(1 - z^{-1})E(z) = \frac{z-1}{Tz}E(z). \tag{2.25}$$

z-transform will be discussed in-depth in chapter 4. Thus, the digital counterpart of the derivative (D) control law is

$$G_C(z) = k_D \frac{z-1}{Tz}. \tag{2.26}$$

Substituting $z = e^{Ts}$ in (2.26), one can show that when $T \to 0$, the digital control law $G_C(z) \to k_D s$, which is the analog version of the D-controller. In general, the choice of the sampling period is very important. The value of T should be sufficiently small so that the digital approximation is adequately valid.

There are number of numerical integration rules that can be used to digitally approximate the integral controller. One popular method of numerically approximating the area under a curve is the trapezoidal integration.

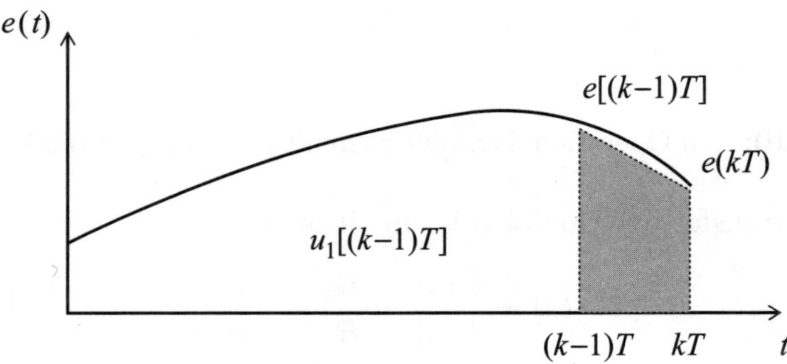

Figure 2.11: Trapezoidal integration.

As shown in Fig. 2.11, the area under $e(t)$ for $(k-1)T \le t < kT$ is approximated by the area of the trapezoid in the interval. And, the integral of $e(t)$ evaluated at $t = kT$ can be obtained as follows. Let

$$u_1(kT) = \int_0^{kT} e(\tau)d\tau. \tag{2.27}$$

Then,

$$u_1(kT) = u_1[(k-1)T] + \frac{T}{2}\left\{e(kT) + e[(k-1)T]\right\}. \tag{2.28}$$

Taking z-transform of the preceding equation enables one to obtain the z-transform of digital integrator as

$$\frac{U_1(z)}{E(z)} = \frac{T(z+1)}{2(z-1)}. \tag{2.29}$$

For integral (I) control, $u(kT) = k_I u_1(kT)$. Hence, the digital counterpart of the integral control law is

$$G_C(z) = \frac{U(z)}{E(z)} = k_I\frac{T(z+1)}{2(z-1)}. \tag{2.30}$$

The above results are sometimes used to obtain an approximate relationship between s and z domains. Such approximations are used for transforming an analog system transfer function, $H(s)$, into a transfer function, $H(z)$, that will describe a discrete system whose behavior approximates the analog system.

Some approximate relationships between s and z domains are as follows. Differentiator and an integrator, respectively, are represented by s and $1/s$ in the s-domain. To that end, we can obtain two approximations to s as follows.

By comparing (2.17) and (2.26), we get

$$s = \frac{(z-1)}{Tz} \quad \text{or} \quad z = 1 + sT. \tag{2.31}$$

By comparing (2.19) and (2.30), we get

$$s = \frac{2(z-1)}{T(z+1)} \quad \text{or} \quad z = \frac{1 + \frac{T}{2}s}{1 - \frac{T}{2}s}. \tag{2.32}$$

Since $z = e^{sT}$ and $s = \frac{1}{T}\ln z$ are unwieldy forms, above approximate transformations are widely used. The result in (2.32) is formally known as *bilinear transformation*.

2.4.1 Ziegler-Nichols Method of Tuning PID Controller

Tuning of a PID controller refers to obtaining the values for the gains, k_P, k_I, and k_D, that best satisfy the design specifications. Ziegler-Nichols method is an empirical method and is widely used in industry. The steps of the tuning procedure are as follows.

1. Recall the PID controller transfer function (2.21)

$$G_C(s) = k_P + k_D s + k_I \frac{1}{s}. \tag{2.33}$$

Set k_I and k_D to be zero and chose a value for k_P for which the system is stable. Now, gradually increase k_P until the system becomes marginally stable. This is when at leat one root gets on the imaginary axis in the s-plane. If the system transfer function is not available, one can observe the system unit step response and reach the situation where step response becomes steadily oscillating.

2. Note the corresponding gain value $k_P = k_{ZN}$ and the frequency of oscillation $\omega_{ZN}[\text{rad}]$.

3. Set the initial values of the PID gains to be

$$k_P = 0.6 k_{ZN}, \quad k_I = \frac{k_{ZN}\omega_{ZN}}{\pi}, \quad \text{and} \quad k_D = \frac{k_{ZN}\pi}{4\omega_{ZN}}. \tag{2.34}$$

4. Fine tune heuristically by gradually decreasing k_I and increasing k_D.

One key feature of this method is that it can be used for complex systems of which the mathematical models are not available. When this is the situation, one can observe the step response on a oscilloscope and read out ω_{ZN} that is direct to obtain when the response is steadily oscillating.

2.5 Frequency-Domain Analysis

Like root locus method, frequency domain analysis where system's frequency response characteristics are taken into account can help predicting and adjusting a system's performance without solving the system differential equations. For many systems such as communication systems, frequency response is of high importance since most of the signals to be processed are composed of sinusoidal components. Further, frequency response analysis is sometimes necessary to evaluate the effect of noise. Noise that can result in poor system performance can be excluded by the design of a passband for the system response. Other key features of frequency response analysis are

- Time-domain performance of the system can be predicted based on the frequency domain characteristics.
- Stability can be analyzed using Nyquist stability criterion that will be discussed shortly.
- Transfer functions of the system components are not necessary as the frequency response can also be determined experimentally by graphically plotting experimental data.

Two graphical methods will be discussed, namely, logarithmic plots (Bode plots) and polar plots (Nyquist's plots). These plots can be used also in the closed-loop design procedure.

2.5.1 Closed-loop Frequency Transfer Function

For the error feedback control system configuration discussed in the preceding chapter, the closed-loop transfer function is

$$M(s) = \frac{Y(s)}{R(s)} = \frac{G(s)}{1 + G(s)H(s)}. \tag{2.35}$$

Under the sinusoidal steady-state, $s = jw$, and the closed-loop frequency transfer function becomes

$$M(j\omega) = \frac{Y(j\omega)}{R(j\omega)} = \frac{G(j\omega)}{1 + G(j\omega)H(j\omega)}. \tag{2.36}$$

The time response can be obtained using inverse Fourier transform as

$$y(t) = \frac{1}{2\pi} \int_{-\infty}^{\infty} Y(j\omega)e^{j\omega t} d\omega. \tag{2.37}$$

If $Y(j\omega)$ is available only as a curve and can not be expressed in analytical form as is often the case in nonlinear system analysis, one can use numerical or graphical integration to obtain the time response, $y(t)$.

Assume that the system is linear. It is known from linear system theory that when the input to a system

$$r(t) = R \sin \omega t \tag{2.38}$$

is sinusoidal with amplitude R and frequency ω, the steady-state output takes the form

$$y(t) = Y \sin(\omega t + \phi) \tag{2.39}$$

with the same frequency but possibly with different amplitude and phase. The angle ϕ is the phase shift.

2.5.2 Frequency-domain Specifications

For a system described by (2.38) and (2.39), the closed-loop frequency response can also be written as

$$\frac{Y(j\omega)}{R(j\omega)} = \frac{G(j\omega)}{1 + G(j\omega)H(j\omega)} = M(\omega)\angle\phi(\omega). \qquad (2.40)$$

For each value of frequency, ω, M is the gain and ϕ is the phase angle between the input and the output. Typical frequency response characteristics of a feedback control system is shown in Fig. 2.12.

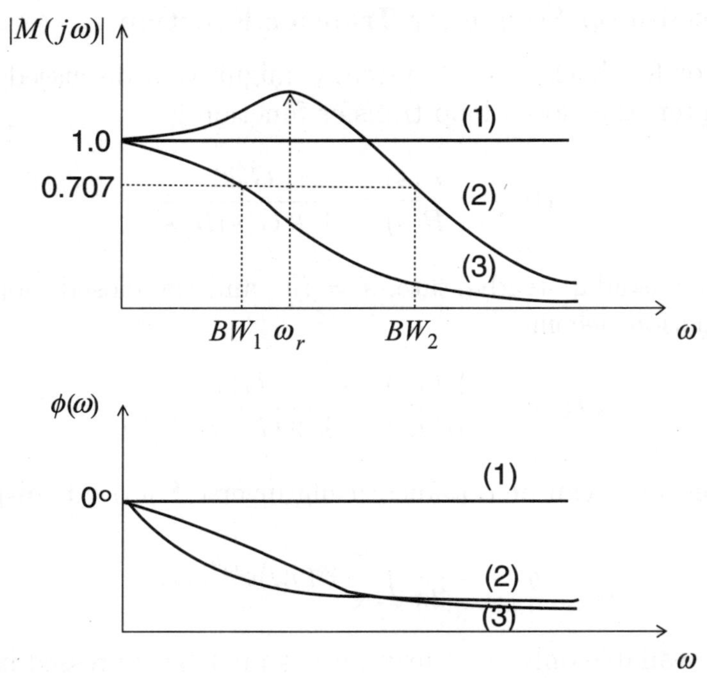

Figure 2.12: Typical gain-phase curves of a feedback control system.

In general, a large resonant peak, M_r, corresponds to a large maximum overshoot of the step response. For most control systems, generally accepted range is $1.1 < M_r < 1.5$. The frequency at which M_r occurs is called resonant frequency, ω_r.

The passband or the bandwidth (BW) is defined as the frequency range up to $M = 0.707$ of the zero frequency value. Spurious noise signals normally occupy a band of frequencies above the frequency band of the true signals. Thus, designing to have a definite passband ensures that the true signals are protected and the noise is attenuated. A larger BW corresponds to a faster rise-time as higher-frequency components of the signal are easily passed through the system. And, smaller BW results in a slow and sluggish time response. Therefore, the BW gives indication of the transient characteristics of the time response. Hence, time-domain performance of the system can be predicted based on the frequency domain characteristics. In frequency domain analysis, the performance specifications are defined in terms of M_r, ω_r, BW, cutoff rate etc. The slope of the log magnitude curve near the cutoff frequency is known as cutoff rate. Two systems can have the same BW but different cutoff rates. In the logarithmic plots (Bode plots) or polar plots (Nyquist's plots) based frequency domain analysis where the system open-loop transfer function (OLTF) is plotted, the specifications are defined in terms of gain margin, phase margin, gain crossover frequency, and phase crossover frequency. These topics will be discussed shortly.

2.5.3 Reshaping the Frequency Response Curves

The effect of adding a pole makes the closed-loop system less stable, while decreasing the band-width. Further,

- Rise-time increases with the decrease of BW
- The larger values of M_r correspond to a larger maximum overshoot in the unit-step response

For example, consider the open-loop transfer function (OLTF) given by $\frac{2}{s(2s+3)}$. Now, adding a pole at $s = -1/b$ will lead to

$$G(s)H(s) = \frac{2}{s(2s+3)(bs+1)}. \tag{2.41}$$

$M(\omega)$ versus ω has been plotted for different locations of the newly added open-loop pole in Fig. 2.13.

In Fig. 2.13, it can be seen that the BW slowly increases with decreasing b. For large values of b, BW decreases while M_r increases sharply. In general, the effect of adding a pole will be a lesser stable closed-loop system with a reduced BW. A sluggish response may be changed to a quicker response by carefully introducing a pole.

The general effect of adding a zero to the OLTF, G(s)H(s), is to increase the BW of the closed-loop system.

For example, a zero at $s = -1/a$ is added to OLTF, $\frac{2}{s(2s+3)}$ to make it $\frac{2(as+1)}{s(2s+3)}$. Same as in the addition of a poles case, it can be shown that over a range of small values of a, the BW is actually increased. For instance, one

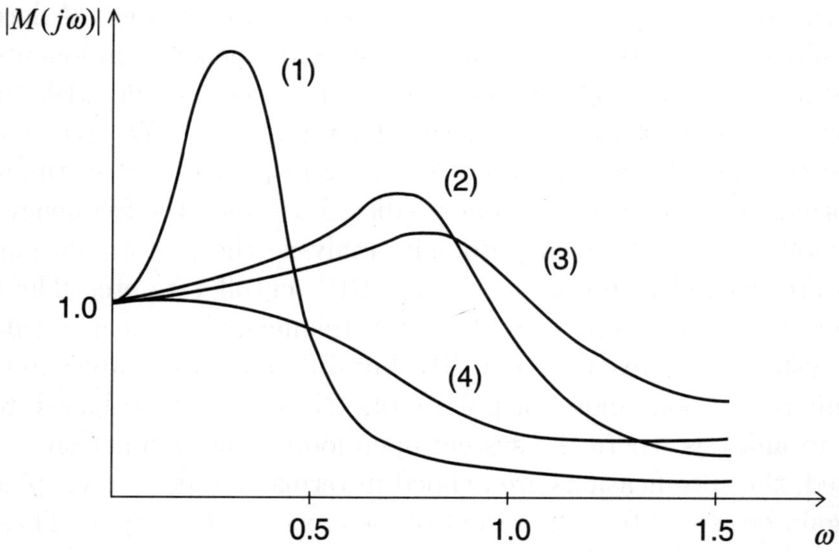

Figure 2.13: Effect of adding a pole at $s = -1/b$: (1) $b = 5$, (2) $b = 1$, (3) $b = 0.5$, (4) $b = 0$.

may better visualize this by plotting the magnitude of $\frac{5(as+1)}{s(5s+2)}$ for different values of a.

2.5.4 Nyquist Stability Criterion

Nyquist criterion is a method that helps determine the stability of a closed-loop system by investigating the properties of frequency domain plots of the open-loop transfer function, $H(s)G(s)$. In the sequel, these plots will be identified as Bode plot and Nyquist plot. The Nyquist criterion has the following features as an alternative option for the design and analysis of control systems.

- It also gives information on the relative stability (degree of stability) giving indication on how to improve the system performance.
- The Nyquist plot of $G(s)H(s)$ gives frequency domain characteristics such as M_r, ω_r, BW etc. with ease.
- Only the OLTF, $G(s)H(s)$, of the system is required. If the OLTF is not available in the analytical form, it may be constructed using experimental data for a open-loop stable system.

Nyquist's stability criteria relates the location of the zeros and pole of $G(s)H(s)$ in s-plane to the frequency response curves of the open-loop transfer function. A rigorous mathematical derivation of this involves complex variable theorem and is available in certain literature.

2.5.5 Bode Plots

Use of a log-scale for the frequency axis expands the low frequency range, which is of primary importance. In logarithmic coordinates, gain/phase versus frequency plots are known as Bode plots. When drawing Bode plots, the open-loop transfer function (OLTF) is used so that Nyquist stability criterion can be applied.

Application of Nyquist stability criterion in terms of log gain/phase diagram is given in Fig. 2.14 and Fig. 2.15. ω_ϕ is the phase margin frequency or gain crossover frequency. This is also the frequency at which the magnitude of $G(j\omega)H(j\omega)$ is unity (0 dB). ω_c is the gain margin frequency or phase crossover frequency. This is also the frequency at which the phase angle is $-180°$.

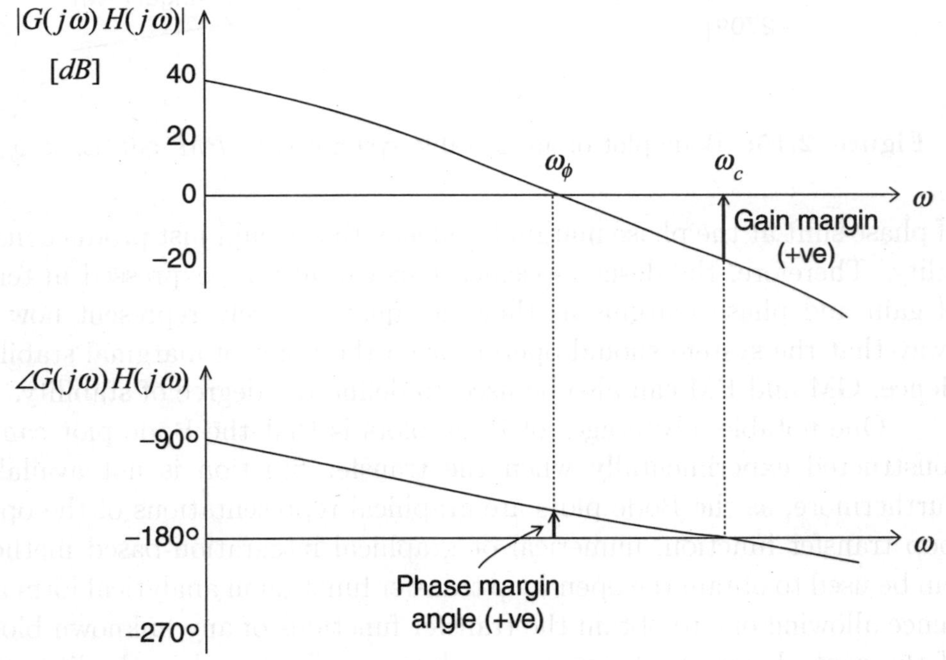

Figure 2.14: Bode plot of a stable system: $GM, PM > 0$, $\omega_c > \omega_\phi$.

The gain margin (GM) gives a qualitative measure on how much the gain must be changed in order to produce either instability in a stable system or to stabilize an unstable system. Phase margin (PM) is the amount

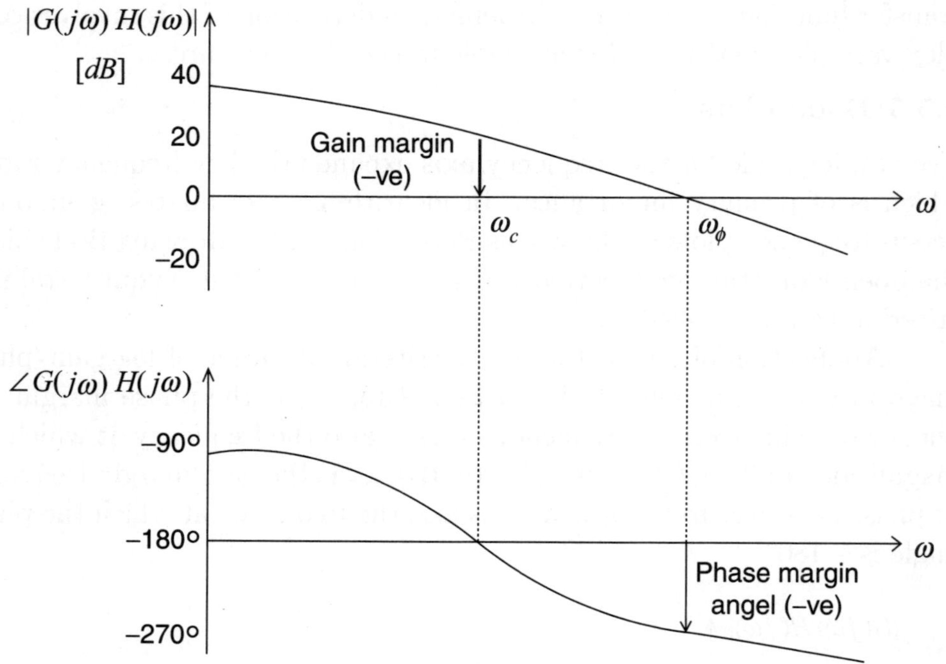

Figure 2.15: Bode plot of an unstable system: $GM, PM < 0$, $\omega_c < \omega_\phi$.

of phase shift at the phase margin frequency that would just produce instability. Therefore, the design specifications can also be expressed in terms of gain and phase margins as they can quantitatively represent how far away that the system should operate from the point of marginal stability. Hence, GM and PM can also be used to define the degree of stability.

One notable advantages of Bode plots is that the Bode plot can be constructed experimentally when the transfer function is not available. Furthermore, as the Bode plots are graphical representations of the open-loop transfer function, numerical or graphical integration-based methods can be used to obtain the open-loop transfer function in analytical form and hence allowing one to obtain the transfer functions of any unknown blocks of the control system. Several procedures can be found in the literature and are omitted in this book.

2.5.5.1 Steps to plot Bode plot   In the absence of a computer, a Bode plot can be sketched by approximating the magnitude and phase with straight line segments. Following steps can be followed to manually construct the Bode plot of a system whose transfer function is $T(s)$.

- Step 1: Factor the numerator and denominator of $T(s)$ and replace s with $j\omega$

- Step 2: Draw the magnitude plot for each factor and add them together to obtain the resultant curve
- Step 3: Draw the phase plot of each factor and add them together to get the resultant curve

To obtain the Bode plot of a transfer function, one can use the Matlab *bode* command. For example

$$bode(50, [1\ 9\ 30\ 40]) \tag{2.42}$$

displays the Bode plot that corresponds to

$$T(s) = \frac{50}{s^3 + 9s^2 + 30s + 40}. \tag{2.43}$$

This is shown in Fig. 2.16 where the frequency is on the logarithmic scale, the phase is given in degrees, and the magnitude is given as the gain in decibels.

$$\text{The gain in decibels (dB) is defined as } 20\log_{10}|T(j\omega)|. \tag{2.44}$$

By right clicking on the magnitude-frequency plot by Matlab, one can obtain the magnitude-frequency coordinate for that point. Likewise, phase-frequency coordinates of any point can be read by clicking on the phase-frequency plot.

Example 2.2: A unity feedback control system has the forward path transfer function given by

$$G(s) = \frac{40}{s^2(s+2)(s+5)}. \tag{2.45}$$

- Step 1: Open-loop frequency transfer function is

$$G(j\omega)H(j\omega) = \frac{4}{(j\omega)^2(1 + j\frac{\omega}{2})(1 + j\frac{\omega}{5})}. \tag{2.46}$$

- Step 2: The magnitude of $|G(j\omega)H(j\omega)|$ in decibels is given by

$$20\log_{10}|G(j\omega)H(j\omega)| = 20\log_{10}4 - 20\log_{10}|(j\omega)^2| - 20\log_{10}|1 + \frac{j\omega}{2}|$$

$$-20\log_{10}|1 + \frac{j\omega}{5}|. \tag{2.47}$$

Let us consider each term separately. Now,

$$20\log 4 = 12 \text{ (dB)} \tag{2.48}$$

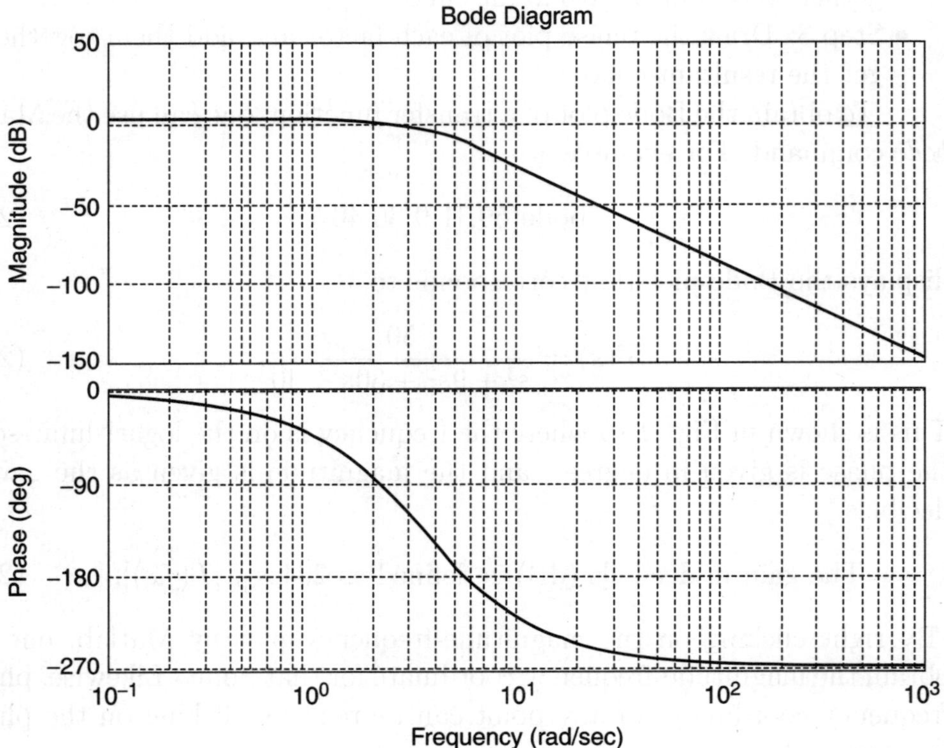

Figure 2.16: Bode plot produced by the Matlab command, $bode(50, [1\ 9\ 30\ 40])$.

is a horizontal line through 12 dB point of the vertical axis.

$$-20 \log |(j\omega)^2| = -40 \log \omega \ \text{(dB)} \tag{2.49}$$

is a straight line having a constant slope of -40 dB/decade and passing through 0 dB point at $\log \omega = 0$ or $\omega = 1$. Now, consider

$$-20 \log_{10} |1 + \frac{j\omega}{2}| = -20 \log \sqrt{1 + \frac{\omega^2}{4}} \ \text{(dB)}. \tag{2.50}$$

An approximate sketch of the preceding equation can be obtained as follows. At low frequencies, $\frac{\omega}{4} << 1$, and hence

$$-20 \log_{10} |1 + \frac{j\omega}{2}| \approx -20 \log 1 = 0 \ \text{(dB)}. \tag{2.51}$$

At high frequencies, $\frac{\omega}{4} >> 1$, and hence

$$-20\log_{10}\left|1 + \frac{j\omega}{2}\right| \approx -20\log\frac{\omega}{2} = -20\log\omega + 20\log 2 \ \text{(dB)}. \quad (2.52)$$

The corner frequency can be obtained by solving $-20\log\frac{\omega_c}{2} = -20\log 1$ $= 0$ that yields $\omega_c = 2$. Therefore, the term $-20\log_{10}\left|1 + \frac{j\omega}{2}\right|$ in (2.47) produces a sketch having corner frequency at $\omega = 2$ and a slope of -20 dB/decade. Likewise, the last term of (2.47) contributes a sketch having corner frequency at $\omega = 5$ and a slope of -20 dB/decade. Now, the magnitude plots of individual terms are added together to obtain the resultant curve. Slope in the resultant curve will be -60 dB/decade for $2 \leq \omega < 5$ and -80 dB/decade for $\omega \geq 5$.

- Step 3: Obtaining the phase angle plot remains to be addressed. From (2.46), it follows that

$$\angle G(j\omega)H(j\omega) = -2 \times 90° - \tan^{-1}\frac{\omega}{2} - \tan^{-1}\frac{\omega}{5}. \quad (2.53)$$

Now, the phase angle plot can be drawn by evaluating the preceding equation at different values of ω $(0 \leq \omega < \infty)$ and connecting them. The complete Bode plots sketched this way is close enough for most practical purposes. The Bode plot produced by Matlab is shown in Fig. 2.17.

2.5.6 Nyquist Plots

Any frequency transfer function $T(j\omega)$ whose magnitude and phase are denoted, respectively, by $|T(j\omega)|$ and $\angle T(j\omega)$ can be plotted on polar coordinates for ω from 0 to ∞. The plots of both $|T(j\omega)|$ and $\angle T(j\omega)$ as a function of ω gives a polar plot. For example, consider the following frequency transfer function

$$T(j\omega) = \frac{1}{(1 + j\omega a)(1 + j\omega b)}. \quad (2.54)$$

The magnitude and phase are given, respectively, by

$$|T(j\omega)| = \frac{1}{\sqrt{(1 - \omega^2 ab)^2 + \omega(a + b)^2}} \quad (2.55)$$

and

$$\angle T(j\omega) = \tan^{-1}\left\{\frac{\omega(a + b)^2}{1 - \omega^2 ab}\right\}. \quad (2.56)$$

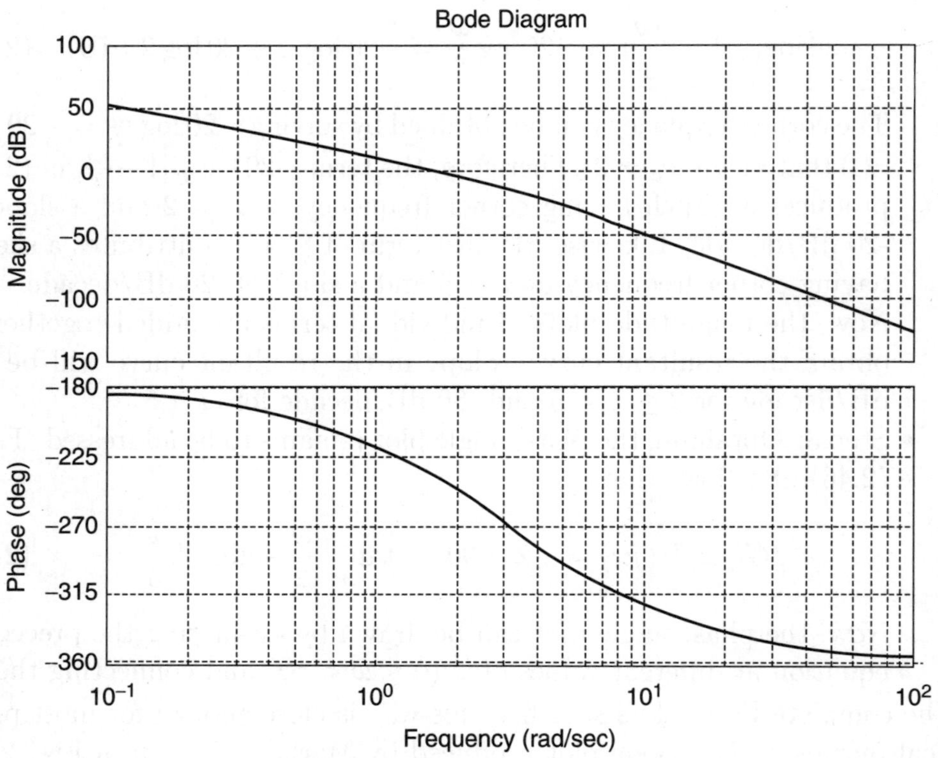

Figure 2.17: Bode plot produced by the Matlab command, bode(40, [1 7 10 0 0]).

$$\text{At} \quad \omega = 0, \quad |T(j\omega)| = 1 \text{ and } \angle T(j\omega) = 0°$$

$$\omega = \infty, \quad |T(j\omega)| = 0 \text{ and } \angle T(j\omega) = -180°$$

$$\omega = \frac{1}{\sqrt{ab}}, \; |T(j\omega)| = \frac{\sqrt{ab}}{a+b} \text{ and } \angle T(j\omega) = -90°.$$

In control applications, the Nyquist plot allows one to predict the stability and performance of a closed-loop system by observing its open-loop behavior. Nyquist plots-based method does not require open-loop stability. Recall that Bode plots methods assume that the system is open loop stable. To that end, Nyquist plots allows one to use Nyquists criterion to determine closed-loop stability when the Bode plots are difficult to be used.

2.5.6.1 Cauchy criterion Nyquist stability criterion is based on the Cauchy criterion from complex analysis. The Cauchy criterion states that when taking a closed contour in the complex plane, and mapping it through a complex function $T(s)$, the number of times that the plot of $T(s)$ encircles the origin is equal to the number of zeros of $T(s)$ enclosed by the frequency contour minus the number of poles of $T(s)$ enclosed by the frequency contour. Encirclements of the origin are counted as positive if they are in the same direction as the original closed contour or negative if they are in the opposite direction.

Consider the closed-loop transfer function, $G(s)/(1 + G(s)H(s))$. If $1 + G(s)H(s)$ encircles the origin, then the OLTF, $G(s)H(s)$, will enclose the point $(-1 + j0)$. Since it is the closed-loop stability that is of interest, we want to know if there are any closed-loop poles or zeros of $1 + G(s)H(s)$ in the right-half plane.

According to the Cauchy criterion, the number N of times that the plot of $G(s)H(s)$ encircles $(-1 + j0)$ is equal to the number Z of zeros of $1 + G(s)H(s)$ enclosed by the closed contour covering entire right-half of s-plane (RHP) minus the number P of poles of $1 + G(s)H(s)$ enclosed by the closed contour $(N = Z - P)$. Based on the Cauchy criterion, the Nyquist criterion then states that

- $P =$ the number of RHP poles of $G(s)H(s)$ as well as of $1 + G(s)H(s)$
- $N =$ the number of times the Nyquist plot of $G(s)H(s)$ encircles $(-1 + j0)$
- $Z =$ the number of RHP poles (unstable) of the closed-loop system

then, $Z = P + N$. Clockwise encirclements of $(-1 + j0)$ are counted as positive encirclements and counter-clockwise encirclements as negative encirclements. Hence, $Z = 0$ and the system will be stable if

- There are no clockwise encirclements about $(-1 + j0)$ and there are no RHP poles
- There are a number of anticlockwise encirclements and an equal number of RHP poles.

Using Nyquist polar plots and Nyquist criterion, it is also possible to find the relative stability of the system. If the Nyquist plot is close to $(-1 + j0)$ point, the system may be on the verge of instability. The proximity to $(-1 + j0)$ point is quantified in terms of the gain margin and the phase margin.

Chapter 3

State-Space Methods

3.1 Introduction

The system formulation in the form of a transfer function is the ratio between the output and the input of the system in the Laplace domain. A transfer function in general is a ratio between two polynomials of s, for a time-invariant system, with constant coefficients. The coefficients can be given as a function of t for a time-varying system. However, this formulation can be too simplistic and inadequate for some systems where the input-output relationship can also be dependent on some other internal variables.

For instance, consider steering control of a motorcycle on a curved road. By experience, it is known that proper steering control (to keep the vehicle on the road centerline while ensuring good ride comfort) and even ensuring stability is comparatively difficult at low and very high speeds. Therefore, the variable *speed* may also be taken into account to realize better control strategies that will be valid over a wider range of speeds. In order to do this, not only the input (steering angle)-output (heading direction), but also the velocity should be included in the system representation. Furthermore, this control problem is better be considered as a control of a multi-input (steering angle and engine/braking torque) multi-output (heading direction and absolute position of the vehicle) or MIMO system. System formulation as a MIMO system also enables one to conveniently include the coupling between the speed control and steering control. This is difficult, if not impossible with transfer function-based system representation, analysis, and design.

Quite in contrast to transfer function-based system formulation, the state-space or state variable method takes into account the internal states of a system as well as its inputs and outputs.

For instance, if the control problem is steering control of a vehicular system, input may be the steering angle and the output the heading

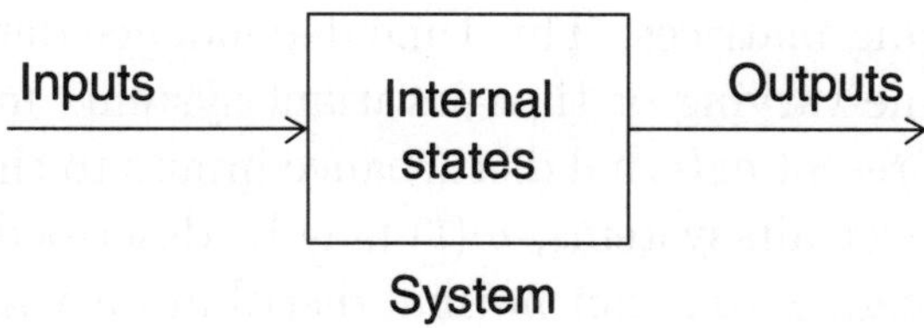

Figure 3.1: Dynamical system with its internal states, inputs and outputs.

direction. The internal states may be the linear speed, yaw rate, and acceleration etc. Taking both internal and external variables, the state-space or state variable method more completely describes a system and provides a more powerful system formulation. State variable methods can apply to both linear and nonlinear systems as well as multi-input multi-output (MIMO) systems.

3.2 State Variable Description

Let $\Re$ denote the real scalars, $\Re^n$ denote the real n-dimensional vectors, and $\Re^{m \times n}$ denote the real $m \times n$ matrices.

State variable description can be given either in the time-domain or in the Laplace domain. By Laplace transforming the time-domain equations, one can obtain the state variable description of a system in the Laplace domain. Let us first consider time domain description. Let the inputs to the system be given by the input vector

$$\boldsymbol{u}(t) = [u_1(t) \ u_2(t) \dots u_p(t)]^T.$$

Likewise, let

$$\boldsymbol{y}(t) = [y_1(t) \ y_2(t) \dots y_q(t)]^T,$$
$$\boldsymbol{x}(t) = [x_1(t) \ x_2(t) \dots x_n(t)]^T$$

respectively, be the vectors of system outputs and the system states. $\boldsymbol{y}(t)$ is also called the measurement vector as it represents q measurements (i.e., measured outputs) of the system. The state variable description of a linear system is

$$\frac{d\boldsymbol{x}(t)}{dt} = \dot{\boldsymbol{x}}(t) = A\boldsymbol{x}(t) + B\boldsymbol{u}(t) + D\boldsymbol{w}(t) \tag{3.1}$$

$$\boldsymbol{y}(t) = C\boldsymbol{x}(t) + \boldsymbol{v}(t). \tag{3.2}$$

Equation (3.1) is a set of first order differential equations stacked together to form a matrix equation. $A \in \Re^{n \times n}$ is an $n \times n$ matrix known as the system matrix. $B \in \Re^{n \times p}$ is system input matrix. A, B, and D are constant or time varying matrices. This time dependence determines whether the system is a time-varying or time-invariant system. $\boldsymbol{w}(t)$ is the system noise that may represent external disturbance inputs to the system, modeling errors etc. For certain systems, $\boldsymbol{w}(t)$ may be described by a probability density function, e.g., zero-mean normal distribution with a known standard deviation. (3.1) is called the *state equation*. State equation describes the rate of change of the states as a function of the states itself and the inputs in the presence of noise.

Equation (3.2) that gives the measured outputs of the system is called the *output equation*. In (3.2), the q measurements of the system have been assumed to be a linear combination of the states, $\boldsymbol{x}(t)$ corrupted with noise. This noise is known as additive noise or measurement noise that is often associated with the sensor. This is often known in terms of the parameters of its probability distribution function. $C \in \Re^{q \times n}$ is the measurement matrix. To that end, (3.2) depends on the sensors that are used.

Equations (3.1) and (3.2) in combination forms the dynamic equations of a system. For instance, if the system is a vehicle or a robot, the system equations (state equation in particular) are called the equations of motion. This is because these equations describe the relation of its motion variables such as speed and acceleration to the deterministic inputs such as steering, engine, braking inputs and disturbance inputs such as external forces due to wind.

Example 3.1: Consider a dc motor with a separately excited field winding. The equivalent electrical circuit is given in Fig. 3.2. In field-weakening control, to obtain speeds beyond its rated value, the steady-state field flux is decreased by decreasing the field current, $I_f(= V_f/R_f)$, where R_f is the resistance of the field winding. In this region also called constant-power region, armature supply voltage, back emf, and current are kept constant at their rated values. As it is often desirable to control the transient behavior, let the instantaneous values of the field voltage and current be given by $v_f(t)$ and $i_f(t)$, respectively. Therefore, in the system formulation, the input to the system (dc motor) may be selected as $v_f(t)$, and depending on the control algorithm and the sensors selected, the output may be selected as the motor speed, $\dot{\theta}(t)$.

As the armature circuit is not involved in this control problem and it is the field circuit that is of interest, Kirchoff voltage law equation may be written for the field circuit as

$$L_f \frac{di_f(t)}{dt} + R_f i_f(t) = v_f(t) \tag{3.3}$$

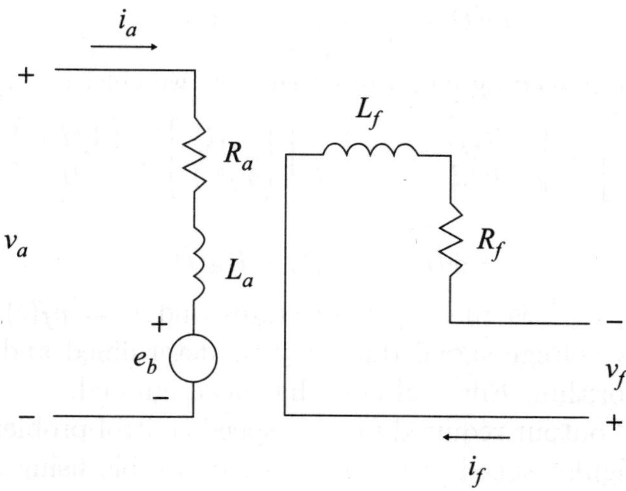

Figure 3.2: Equivalent circuit of a separately exited dc motor.

where L_f is the field winding inductance. Assume an inertial mechanical load of inertia J. Let the friction coefficient be f. Electromagnetic torque produced by the motor due to interaction of the field flux, ϕ_f, and the armature current, i_a, is

$$T_m(t) = k_t \phi_f(t) i_a(t)$$

where k_t is the torque constant of the motor. To that end, for the underlining constant-power region, the motor torque may be approximated as $T_m(t) = k i_f(t)$ where k is a constant. Let the motor shaft angular position be $\theta(t)$. Now, the equation of the mechanical subsystem can be written as

$$J\frac{d^2\theta(t)}{dt^2} + f\frac{d\theta(t)}{dt} = k i_f(t). \qquad (3.4)$$

Recall that the state equation is a set of first order differential equations. In order to obtain the state variable description, the states may now be defined as

$$x_1(t) = i_f(t)$$
$$x_2(t) = \dot{\theta}(t).$$

Now, (3.3) and (3.4) becomes

$$\dot{x}_1(t) = -\frac{R_f}{L_f}x_1(t) + \frac{1}{L_f}v_f(t)$$

$$\dot{x}_2(t) = \frac{k}{J}x_1(t) - \frac{f}{J}x_2(t).$$

By stacking the preceding equations together, we get

$$\begin{bmatrix} \dot{x}_1(t) \\ \dot{x}_2(t) \end{bmatrix} = \begin{bmatrix} -R_f/L_f & 0 \\ k/J & -f/J \end{bmatrix} \begin{bmatrix} x_1(t) \\ x_2(t) \end{bmatrix} + \begin{bmatrix} 1/L_f \\ 0 \end{bmatrix} v_f(t). \qquad (3.5)$$

or

$$\dot{\boldsymbol{x}}(t) = A\boldsymbol{x}(t) + Bu(t) \qquad (3.6)$$

where $\boldsymbol{x} = [x_1 \; x_2]^T$ is the vector of states and $u = v_f(t)$ is the input. Input here is a voltage signal that may be determined and produced by the control algorithm. Effect of noise has been ignored.

Measured output required for this speed control problem may be the motor shaft angular speed, $\dot{\theta}(t)$. This is measurable using a tachometer. To this end, the output equation may be chosen as

$$\dot{\theta}(t) = \begin{bmatrix} 0 & 1 \end{bmatrix} \begin{bmatrix} x_1(t) \\ x_2(t) \end{bmatrix} \qquad (3.7)$$

or

$$y(t) = C\boldsymbol{x}(t). \qquad (3.8)$$

Here also, measurement noise has been ignored. It is noteworthy that the state space representation in the Laplace domain can simply be obtained by taking the Laplace transform of (3.6) and (3.8). Since this is a single-input single-output (SISO) system, the transfer function of the system, $\frac{\Theta(s)}{V_f(s)}$, can be obtain by solving the Laplace transforms of (3.3) and (3.4) neglecting the initial conditions.

The state variable representation (3.6) and (3.8) can be used for analysis and design of control systems to realize speed control of the motor. For instance, refer stability analysis, concepts of controllability and observability that follow in this chapter. The block diagram of a standard error feedback control system for this application is given in Fig. 3.3 where $\dot{\theta}_d(t)$ is the reference input or the desired speed.

Example 3.2: If in contrast to example 4.1, the motor position is to be controlled, the controlled variable will be the angular position, $\theta(t)$, of the motor shaft. In order to obtain a state variable description that can be used to realize a position control system, the states may be defined differently as

$$x_1(t) = i_f(t)$$
$$x_2(t) = \theta(t)$$
$$x_3(t) = \dot{\theta}(t).$$

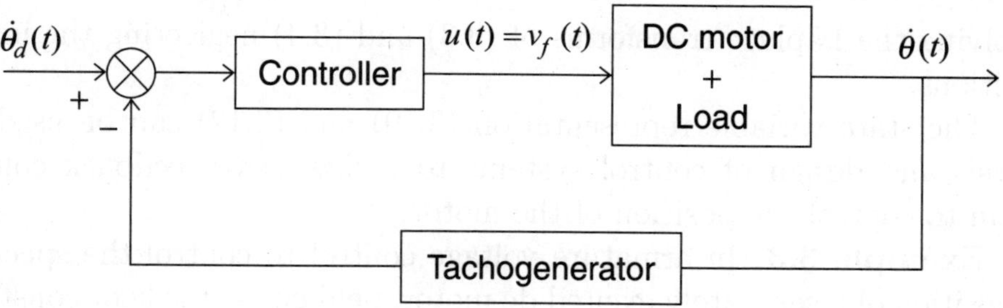

Figure 3.3: DC motor speed control system.

Now, (3.3) and (3.4) becomes

$$\dot{x}_1(t) = -\frac{R_f}{L_f}x_1(t) + \frac{1}{L_f}v_f(t)$$
$$\dot{x}_2(t) = x_3(t)$$
$$\dot{x}_3(t) = \frac{k}{J}x_1(t) - \frac{f}{J}x_3(t).$$

By stacking the preceding equations together, we get

$$\begin{bmatrix} \dot{x}_1(t) \\ \dot{x}_2(t) \\ \dot{x}_3(t) \end{bmatrix} = \begin{bmatrix} -R_f/L_f & 0 & 0 \\ 0 & 0 & 1 \\ k/J & 0 & -f/J \end{bmatrix} \begin{bmatrix} x_1(t) \\ x_2(t) \\ x_3(t) \end{bmatrix} + \begin{bmatrix} 1/L_f \\ 0 \\ 0 \end{bmatrix} v_f(t). \quad (3.9)$$

or

$$\dot{\boldsymbol{x}}(t) = A\boldsymbol{x}(t) + Bu(t) \quad (3.10)$$

where $\boldsymbol{x} = [x_1\ x_2\ x_3]^T$ is the vector of states and $u = v_f(t)$ is the input.

Measured output required for this speed control problem may be the motor shaft angular position, $\theta(t)$. This is usually measurable using an optical encoder. To this end, the output equation may be chosen as

$$\dot{\theta}(t) = \begin{bmatrix} 0 & 1 & 0 \end{bmatrix} \begin{bmatrix} x_1(t) \\ x_2(t) \\ x_3(t) \end{bmatrix} \quad (3.11)$$

or

$$y(t) = C\boldsymbol{x}(t). \quad (3.12)$$

Here also, measurement noise has been ignored. The state space representation in the Laplace domain can simply be obtained by taking the Laplace transform of (3.10) and (3.12). Since this is a single-input single-output

(SISO) system, the transfer function of the system, $\frac{\Theta(s)}{V_f(s)}$, can be obtain by solving the Laplace transforms of (3.3) and (3.4) neglecting the initial conditions.

The state variable representation (3.10) and (3.12) can be used for analysis and design of control systems to realize error feedback control system to control the position of the motor.

Example 3.3: In armature voltage control to control the speed or the position of a separately excited dc motor, field current is kept constant. Since it is the armature voltage that is varied, the Kirchoff's equation may now be written for the armature equivalent circuit as

$$L_a \frac{di_a(t)}{dt} + R_a i_a(t) + e_b = v_a(t) \tag{3.13}$$

where the subscript, a, denotes the quantities associated with the armature and $e_b = k_b \dot{\theta}$ is the back emf with k_b being the back emf constant.

For the same type of load, the equation for the mechanical subsystem remains the same. Since the air gap flux is constant, the electromagnetic torque developed by the motor is proportional to the armature current, i_a.

$$T_m(t) = k_t i_a(t)$$

where k_t is the torque constant of the motor. Now, the equation of the mechanical subsystem is

$$J \frac{d^2\theta(t)}{dt^2} + f \frac{d\theta(t)}{dt} = k_t i_a(t). \tag{3.14}$$

Having speed control in mind, the states may now be defined as

$$x_1(t) = i_a(t)$$
$$x_2(t) = \dot{\theta}(t).$$

Now, (3.13) and (3.14) becomes

$$\dot{x}_1(t) = -\frac{R_a}{L_a} x_1(t) - \frac{k_b}{L_a} x_2(t) + \frac{1}{L_a} v_a(t)$$
$$\dot{x}_2(t) = \frac{k_t}{J} x_1(t) - \frac{f}{J} x_2(t)$$

whence put into the matrix form, we get

$$\begin{bmatrix} \dot{x}_1(t) \\ \dot{x}_2(t) \end{bmatrix} = \begin{bmatrix} -R_a/L_a & -k_b/L_a \\ k_t/J & -f/J \end{bmatrix} \begin{bmatrix} x_1(t) \\ x_2(t) \end{bmatrix} + \begin{bmatrix} 1/L_a \\ 0 \end{bmatrix} v_a(t).$$

or

$$\dot{x}(t) = Ax(t) + Bu(t)$$

where $x = [x_1\ x_2]^T$ is the vector of states and armature voltage supply and $u = v_a(t)$ is the input. The output equation may be chosen as

$$y(t) = Cx(t).$$

where $C = [0\ 1]$.

Defining the states as $x = [x_1(t)\ x_2(t)\ x_3(t)]^T = [i_a(t)\ \theta(t)\ \dot{\theta}(t)]^T$, one may obtain a state-space model suitable for the position control problem of the motor.

One of the major advantage of state-variable methods is their ability to better formulating and analyzing multiple-input multiple-output (MIMO) systems. For convenience, in the sections that follow, we consider for the most part single-input single-output (SISO) systems. It is important to note that the results that will be obtained will also be valid for the MIMO systems.

3.3 Solution of The State Equation

For convenience, consider the following single-input single-output (SISO) system.

$$\dot{x}(t) = Ax(t) + Bu(t) \tag{3.15}$$
$$y(t) = Cx(t) \tag{3.16}$$

where $x(t) \in \Re^n$ is the vector of states, and the scalars $u(t)$ and $y(t)$ are the input and the output, respectively. State equation (3.15) is a first order matrix differential equation and hence can be solved for $x(t)$. Once the states, $x(t)$, are known, the output, $y(t)$, which is a linear combination of the states can be obtained.

The general solution of (3.15) contains two parts, namely, the zero-state response and the zero-input response. Zero-state response is the part of the response (solution) purely due to the external input, $u(t)$. Zero-input response is the part of the response purely due to the initial conditions. This can be explained taking a practical example as follows. Consider a moving vehicle. The variation of its internal states such as velocity components, yaw rate etc will depend on the engine-braking torque input and the steering input as well as the initial conditions such as the starting velocity and heading. Measured responses such as position, resultant velocity, heading direction etc are dependent on the internal states. For instance, if the control inputs are all made zero while a ground vehicle is moving on a flat road, its states will keep changing purely due to the initial conditions. It will keep moving due to the momentum (caused by the velocity) that it has. On the other hand, a vehicle starting from the rest will move purely due to the input torque. If the system equations (3.15) are solved for the internal states, $\boldsymbol{x}(t)$ for $t \geq t_0$, the general solution will contain a component purely due to $u(t)$ and another purely due to $\boldsymbol{x}(t_0)$. Often, t_0 is chosen as $t_0 = 0$.

Taking the Laplace transform of (3.15), we get

$$s\boldsymbol{X}(s) - \boldsymbol{x}(0) = A\boldsymbol{X}(s) + BU(s) \tag{3.17}$$

where $\boldsymbol{x}(t) \leftrightarrow \boldsymbol{X}(s)$ and $u(t) \leftrightarrow U(s)$ are Laplace transform pairs and $\boldsymbol{x}(0) = \boldsymbol{x}(t = 0)$ is the vector of initial conditions.

$$\boldsymbol{X}(s) = [X_1(s) \ X_2(s) \ldots X_n(s)]^T$$
$$\boldsymbol{x}(0) = [x_1(0) \ x_2(0) \ldots x_n(0)]^T.$$

Rearranging the terms in the preceding equation yields

$$[sI - A]\boldsymbol{X}(s) = BU(s) + \boldsymbol{x}(0) \tag{3.18}$$

that results in

$$\boldsymbol{X}(s) = [sI - A]^{-1}BU(s) + [sI - A]^{-1}\boldsymbol{x}(0). \tag{3.19}$$

Assume that $[sI - A]^{-1}$ is nonsingular. Taking the inverse Laplace transform yields

$$\boldsymbol{x}(t) = \mathcal{L}^{-1}[(sI - A)^{-1}BU(s)] + [\mathcal{L}^{-1}(sI - A)^{-1}]\boldsymbol{x}(0) \qquad t \geq 0. \tag{3.20}$$

This is known as the state-transition equation. Now, $y(t)$ can be obtained from (3.16) as

$$y(t) = C\mathcal{L}^{-1}[(sI - A)^{-1}BU(s)] + C[\mathcal{L}^{-1}(sI - A)^{-1}]\boldsymbol{x}(0) \quad t \geq 0. \tag{3.21}$$

The first term on the right-hand side of (3.21) is the zero-state response and the last term is the zero-input response.

Example 3.4: Consider a system, which is characterized by

$$\dot{x}_1(t) = 0$$
$$\dot{x}_2(t) = 30x_1(t) - 30x_2(t).$$

Note that no external input is present in the equations. Hence, by solving these state equations, one will only obtain the free response of the system. This is the response that is excited by the initial conditions only.

Once the above equations are put into the matrix form, we get

$$A = \begin{bmatrix} 0 & 0 \\ 30 & -30 \end{bmatrix} \quad \text{and} \quad sI - A = \begin{bmatrix} s & 0 \\ -30 & s+30 \end{bmatrix}.$$

Therefore,

$$[sI - A]^{-1} = \frac{1}{s(s+30)} \begin{bmatrix} s+30 & 0 \\ 30 & s \end{bmatrix} = \begin{bmatrix} 1/s & 0 \\ a/s(s+30) & 1/(s+30) \end{bmatrix}$$

$$= \frac{1}{30s} \begin{bmatrix} 30 & 0 \\ 30 & 0 \end{bmatrix} - \frac{1}{30(s+30)} \begin{bmatrix} 0 & 0 \\ 30 & -30 \end{bmatrix}.$$

Zero-input response can be found using

$$\boldsymbol{x}(t) = \mathcal{L}^{-1}[(sI - A)^{-1}]\boldsymbol{x}(0)$$
$$\begin{bmatrix} x_1(t) \\ x_2(t) \end{bmatrix} = \begin{bmatrix} 1 & 0 \\ 1 - e^{-30t} & e^{-30t} \end{bmatrix} \begin{bmatrix} x_1(0) \\ x_2(0) \end{bmatrix}$$

which can be evaluated if the values of $x_1(0)$ and $x_2(0)$ are known. It is direct to obtain the measured response, $y(t)$, if the output equation is known.

3.3.1 State-transition Matrix

State-transition matrix enables write the total system response in the time domain allowing the system to be analyzed directly in the time domain. Even though (3.20) allows writing the total time response, it is sometimes useful to write the state-transition equation exclusively using time domain terms. Besides, state-transition equation in (3.20) is useful only when the initial time is defined to be at $t = 0$. Especially, in the study of discrete-time control systems, it is often desirable to break up the transition process of the states into a sequence of transitions. Therefore, a more general initial time, say t_0, may be desirable in the state-transition equation.

In order to obtain this result, note that

$$\frac{d}{dt}[e^{-At}\boldsymbol{x}(t)] = e^{-At}[\dot{\boldsymbol{x}}(t) - A\boldsymbol{x}(t)] \tag{3.22}$$

where A is the system matrix in (3.15). Substituting for the right-hand side of (3.22) from (3.15), we get

$$\frac{d}{dt}[e^{-At}\boldsymbol{x}(t)] = e^{-At}Bu(t). \tag{3.23}$$

Integrating both sides yields

$$\int_{t_0}^{t} d\left(e^{-At}\boldsymbol{x}(t)\right) = \int_{t_0}^{t} e^{-A\tau}Bu(\tau)d\tau \tag{3.24}$$

which after some manipulation becomes

$$\boldsymbol{x}(t) = e^{A(t-t_0)}\boldsymbol{x}(t_0) + \int_{t_0}^{t} e^{A(t-\tau)}Bu(\tau)d\tau. \tag{3.25}$$

The state-transition matrix is defined as

$$\phi(t) = e^{At}. \tag{3.26}$$

Writing (3.25) in terms of state-transition matrix gives the transition equation as

$$\boldsymbol{x}(t) = \phi(t - t_0)\boldsymbol{x}(t_0) + \int_{t_0}^{t} \phi(t - \tau)Bu(\tau)d\tau. \tag{3.27}$$

The following property of the state-transition matrix is noteworthy.

$$\begin{aligned}
\phi(t_n - t_{n-1}) \cdots \phi(t_2 - t_1)\phi(t_1 - t_0) &= e^{A(t_n-t_{n-1})} \ldots e^{A(t_2-t_1)}e^{A(t_1-t_0)} \\
&= e^{A(t_n-t_0)} = \phi(t_n - t_0)
\end{aligned} \tag{3.28}$$

for any $t_0, t_1, \cdots t_n$. This property implies that the transition process of the states can be divided into a series of consecutive transitions. There are several methods that may be used to determine the state-transition matrix.

- It is apparent that (3.25) reverts to (3.20) when $t_0 = 0$. By comparing the two equations, the state-transition matrix may be obtained as

$$\phi(t) = \mathcal{L}^{-1}[(sI - A)^{-1}]. \tag{3.29}$$

- By evaluating the infinite series

$$\phi(t) = e^{At} = I_{n\times n} + tA + \frac{t^2}{2!} + \frac{t^3}{3!} + \cdots. \tag{3.30}$$

3.3.2 Characteristic Equation and the Eigenvalues

We already know that the characteristic equation plays an important role in the study of linear systems. If the transfer function is available, characteristic equation is obtained by equating the denominator polynomial to zero. Characteristic equation can also be defined with respect to the state equations.

A solution like (3.21) is only possible if the equation (3.19) exists to be inverse Laplace transformed. $(sI - A)^{-1}$ in (3.19) is known as the resolvent matrix and can also be written as

$$(sI - A)^{-1} = \Phi(s) = \frac{\text{adj}(sI - A)}{|sI - A|}. \tag{3.31}$$

Substituting (3.19) in the Laplace transform of (3.16) yields

$$Y(s) = C[sI - A]^{-1}BU(s) + C[sI - A]^{-1}\boldsymbol{x}(0) \tag{3.32}$$

or

$$Y(s) = C\frac{\text{adj}(sI - A)}{|sI - A|}BU(s) + C\frac{\text{adj}(sI - A)}{|sI - A|}\boldsymbol{x}(0). \tag{3.33}$$

Setting the denominator of the preceding equation to zero, we get the characteristics equation as

$$|sI - A| = 0. \tag{3.34}$$

(3.34) is an alternative form of the characteristic equation and should yield the same results as in transfer function analysis. Recall that by definition, transfer function does not include the initial conditions. The transfer function, $\frac{Y(s)}{U(s)}$, of a SISO system can be obtained by ignoring the zero-input response in (3.33). Characteristic equation is obtained by equating the denominator polynomial, $|sI - A|$, to zero leading to the same result.

An important property of the characteristic equation is that if the coefficients of A is real, the coefficients of $|sI - A|$ will also be real. The roots of the characteristic equation, $|sI - A| = 0$, are also referred to as the *eigenvalues* of the matrix A.

3.3.3 Stability and the Eigenvalues

Recall from (3.29) and (3.31) that the state-transition matrix and the resolvent matrix is a Laplace transform pair, $\phi(t) \leftrightarrow \Phi(s)$. Let the roots of the characteristic equation $|sI - A| = 0$ be $s_1, s_2, \cdots s_n$. Now, $\Phi(s)$ in (3.31) may be written as

$$\Phi(s) = \frac{K_1}{s - s_1} + \frac{K_2}{s - s_2} + \cdots + \frac{K_n}{s - s_n} \tag{3.35}$$

where K_i; $i = 1, 2, \cdots n$ are matrices that can be found using partial fractions as

$$K_i = \text{Lim}_{s \to s_i} \left\{ [sI - A]^{-1}(s - s_i) \right\}. \tag{3.36}$$

State-transition matrix, if the roots are distinct or not repeating, will take the form

$$\mathcal{L}^{-1}[\Phi(s)] = \phi(t) = K_1 e^{s_1 t} + K_2 e^{s_2 t} + \cdots + K_n e^{s_n t} \tag{3.37}$$

and if there are repeating roots

$$\phi(t) = K_1 e^{s_1 t} + K_2 e^{s_2 t} + K_3 t e^{s_2 t} + \cdots + K_n e^{s_n t}. \tag{3.38}$$

According to the above results, for the system response to be stable, roots of the characteristics equation or eigenvalues of A should not have positive real parts. Consider the terms due to the repeating root, s_2, in (3.38). If this root is purely imaginary, the term

- $K_2 e^{s_2 t}$ will be steadily oscillating and hence marginally stable
- $K_2 t e^{s_2 t}$ will be unstable as $t \to \infty$.

Hence, if the roots are repeating, they can not be on the imaginary axis that includes the origin of the s-plane.

It is noteworthy that even though a SISO system was considered in the various analyzes in the preceding sections to avoid unwieldy formulations, the results obtained will hold also for MIMO systems.

Example 3.5: In the attitude control systems of satellites, small rocket thrusters are used as the actuators to keep the satellite in the correct angular positions to point its antennas (in communication satellites) and instruments (such as telescopes in astronomical satellites) in the right directions. Solar panels need to be oriented towards the sun for maximum power generation. Gyroscopes, sun sensors, star trackers, and horizon indicators are the most common sensors used to measure the attitude angles; roll, pitch, and yaw. Torques applied by firing thrusters re-orient the vehicle to a desired attitude. Dynamic equations can be obtained by writing Euler's equation (net torque equals the rate of change of angular momentum) for angular motions about each of the 3-axes.

Let θ, ϕ, and ψ be the role, pitch, and yaw angle, respectively. If the nonlinearities can be considered to be mild, dynamic equations can be linearized. Linearized dynamic equations for roll, pitch, and yaw motions are

$$I_x \ddot{\theta}(\iota) + 4(I_y - I_z)\theta(\iota) - (I_y - I_x - I_z)\dot{\psi}(\iota) = \tau_x(\iota) \tag{3.39}$$

$$I_y \ddot{\phi}(\iota) + 3(I_x - I_z)\phi(\iota) \qquad\qquad\qquad = \tau_y(\iota) \tag{3.40}$$

$$I_z \ddot{\psi}(\iota) + (I_y - I_x)\psi(\iota) + (I_y - I_x - I_z)\dot{\theta}(\iota) = \tau_z(\iota) \tag{3.41}$$

where $\iota = \omega t$ with ω being the orbital velocity. τ_x, τ_y, and τ_z are the torques applied about the roll, pitch, and yaw axes (x, y, and z), respectively. Principle moments of inertia about the 3 axes are $I_x = 100 \text{ kgm}^2, I_y = 200 \text{ kgm}^2$, and $I_z = 50 \text{ kgm}^2$.

The pitch dynamics given by (3.40) only contains the pitch angle and its time derivatives. Further, since the 3 axes are mutually orthogonal, each of the equations only contains the torque applied about its own axis. Therefore, the pitch control is completely independent of other two control problems, roll and yaw control, and hence can separately be addressed.

Quite in contrast, roll dynamics (3.39) is dependent on the yaw rate, $\dot{\psi}(t)$. Similarly, yaw dynamics (3.41) is dependent on the rolling rate, $\dot{\theta}(t)$. In control engineering terminology, the roll and yaw dynamics are said to be coupled. One of the major advantages of state-space-based system formulation is that the coupling effects like these can be accommodated. Since an equation of the kind of (3.40) can also be analyzed using the methods discussed in the preceding chapters (e.g., transfer function methods), let us consider the coupled dynamics, roll and yaw.

For simplicity, assume that $\tau_x(\iota) = 0$. Now, (3.39) and (3.41) become

$$\ddot{\theta}(\iota) + 6\theta(\iota) - 0.5\dot{\psi}(\iota) = 0$$

$$\ddot{\psi}(\iota) + 2\psi(\iota) + \dot{\theta}(\iota) = \frac{\tau_z(\iota)}{I_z}.$$

Define the state-variables as

$$x_1(\iota) = \theta(\iota), \ \ x_2(\iota) = \dot{\theta}(\iota), \ \ x_3(\iota) = \psi(\iota), \ and \ x_4(\iota) = \dot{\psi}(\iota).$$

Now, the two preceding equations can be re-written as

$$\dot{\boldsymbol{x}}(\iota) = A\boldsymbol{x}(\iota) + Bu(\iota)$$

where

$$A = \begin{bmatrix} 0 & 1 & 0 & 0 \\ -6 & 0 & 0 & 0.5 \\ 0 & 0 & 0 & 1 \\ 0 & -1 & -2 & 0 \end{bmatrix}, \ B = \begin{bmatrix} 0 \\ 0 \\ 0 \\ 1 \end{bmatrix}, \ \text{and} \ u(\iota) = \frac{\tau_z(\iota)}{I_z}.$$

Let us define the output as follows

$$\theta(\iota) = x_1(\iota) = C_1\boldsymbol{x}(\iota)$$
$$\psi(\iota) = x_3(\iota) = C_2\boldsymbol{x}(\iota)$$

where $C_1 = [1\ 0\ 0\ 0]$, $C_2 = [0\ 0\ 1\ 0]$. The characteristic equation is

$$|sI - A| = 0$$
$$2s^4 + 17s^2 + 24 = 0.$$

The roots are $s = \pm 2.6$, $s = \pm 1.3$. Hence, the system will be marginally stable.

$$\boldsymbol{x}(\iota) = \mathcal{L}^{-1}\left\{\frac{\text{adj}(sI - A)}{|sI - A|}BU(s)\right\} + \mathcal{L}^{-1}\left\{\frac{\text{adj}(sI - A)}{|sI - A|}\right\}\boldsymbol{x}(0) \qquad \iota \geq 0.$$

Let us obtain the time response of $\theta(\iota)$. The results for $\psi(\iota)$ can be obtained following similar lines if the input, $\tau_z(\iota)$ is known. Since $\tau_x(\iota) = 0$, considering the zero input response alone is sufficient to obtain $\theta(\iota)$.

$$\text{adj}(sI - A) = \begin{bmatrix} s(s^2 + 2.5) + 0.5s & s^2 + 2 & -1 & 1.5s \\ -6(s^2 + 2) & s(s^2 + 2) & -s & 0.5s^2 \\ 6 & -s & s(s^2 + 6.5) + 6s & s^2 + 6 \\ 6s & s^2 & -2(s^2 + 6) & s(s^2 + 6) \end{bmatrix}.$$

$$\theta(\iota) = x_1(\iota)$$
$$= \mathcal{L}^{-1}\left\{\frac{[s(s^2 + 2) + 0.5s]x_1(0) + (s^2 + 2)x_2(0) - x_3(0) + 0.5sx_4(0)}{(s - j2.6)(s + j2.6)(s - j1.3)(s + j1.3)}\right\}.$$

For simplicity, assume that $x_1(0) = x_2(0) = x_3(0) = 0$. Then

$$\theta(\iota) = 0.5x_4(0)[k_1 \cos 2.6\iota + k_2 \sin 1.3\iota]$$

where k_1, k_2 are constants and $x_4(0) = \dot{\psi}(0)$. It is noteworthy that the coupling between roll and yaw dynamics is also evident in the above solution for $\theta(\iota)$. Variation of the roll angle, $\theta(\iota)$, depends also on the initial yaw rate, $\dot{\psi}(0)$. Hence, $\tau_z(\iota)$ has an influence on roll dynamics as well. A control problem like this can not sufficiently be addressed by the use of transfer function-based methods.

3.4 Controllability and Observability

The concept of controllability and observability are important to understand many aspects of dynamical systems and the behavior of state-space trajectories when the system is excited by a defined class of inputs.

Again for simplicity, we consider a single-input single-output (SISO) system. But the results that will be discussed will also be valid for MIMO systems.

3.4.1 Controllability

Controllability test can help find out what conditions a system needs to satisfy to ensure that its states can be steered to any desired final state, $\boldsymbol{x}(t_f)$, from an initial state, $\boldsymbol{x}(t_0)$, by some unconstrained control, $u(t)$, in a finite time interval, $t_f - t_0$. See Fig. 3.4.

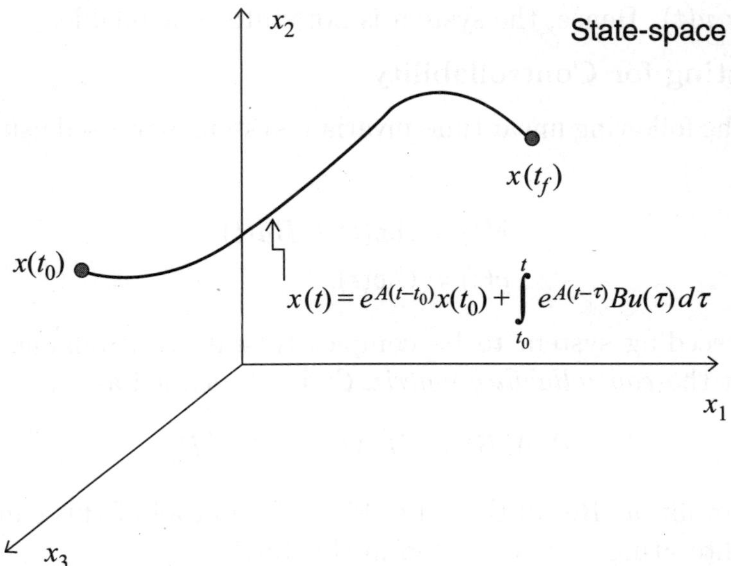

Figure 3.4: State space trajectory.

An example is a vehicular system that is being controlled to reach a particular location and finally oriented in a particular direction. A turbo generator starting from the rest and being controlled to reach a given set of operating conditions such as the terminal voltage, frequency, and the output power. In modern control, quite in contrast to classical control dominated by trial-and-error methods, controllability and observability tests allow the designer to know at the outset if the design solution exists or not for the given system parameters and design specifications. For instance, if any of the state variables is independent of the control input, $u(t)$, it is impossible to steer this particular state to a desired state by applying a suitable control effort. Hence, such a state variable is uncontrollable.

A system is said to be completely state controllable, for any time, t_0, if each initial state, $\boldsymbol{x}(t_0)$, can be steered to any final state, $\boldsymbol{x}(t_f)$, by a permissible unconstrained input, $u(t)$, in a finite time, $t_f - t_0$. Consider, for instance, the following linear system.

$$\dot{x}_1 = -2x_1 + x_2 + u(t) \tag{3.42}$$
$$\dot{x}_2 = -x_2 \tag{3.43}$$

Since the control, $u(t)$, affects only the state $x_1(t)$, it is impossible to steer state $x_2(t)$ from an initial value to a desired final value in a finite time interval by $u(t)$. Hence, the system is not state controllable.

3.4.2 Testing for Controllability

Consider the following linear time-invariant system expressed using familiar notations

$$\dot{\boldsymbol{x}}(t) = A\boldsymbol{x}(t) + Bu(t) \tag{3.44}$$
$$y(t) = C\boldsymbol{x}(t). \tag{3.45}$$

For the preceding system to be completely state controllable, it can be shown that the *controllability matrix, $C(A, B)$*, defined as

$$C(A, B) = [B \ AB \ \dots A^{n-1}B] \tag{3.46}$$

must have rank, n. Recall that $A \in \Re^{n \times n}$. The proof of this can be found in certain literature and is omitted in this book.

Furthermore, controllability of a system only depends on A and B, and is independent of the output matrix, C. The controllability of the system is sometimes referred to as the controllability of the pair $\{A, B\}$.

Example 3.6: Consider the linear system given by (3.42) and (3.43).

$$A = \begin{bmatrix} -2 & 1 \\ 0 & -1 \end{bmatrix}, \quad B = \begin{bmatrix} 1 \\ 0 \end{bmatrix}, \quad \text{and} \quad AB = \begin{bmatrix} -2 & 1 \\ 0 & -1 \end{bmatrix} \begin{bmatrix} 1 \\ 0 \end{bmatrix} = \begin{bmatrix} -2 \\ 0 \end{bmatrix}.$$

For this system, $n = 2$. Hence, the controllability matrix is given by

$$C(A, B) = [B \ AB] = \begin{bmatrix} 1 & -2 \\ 0 & 0 \end{bmatrix}$$

and has rank 1 instead of rank 2 that is required for controllability. In other words, $C(A, B)$ is singular. System is uncontrollable.

Example 3.7: Consider the linear system given by

$$\dot{x}_1 = x_2$$
$$\dot{x}_2 = x_3$$
$$\dot{x}_3 = -12x_2 - 7x_3 + u(t).$$

For this system

$$A = \begin{bmatrix} 0 & 1 & 0 \\ 0 & 0 & 1 \\ 0 & -12 & -7 \end{bmatrix} \quad \text{and} \quad B = \begin{bmatrix} 0 \\ 0 \\ 1 \end{bmatrix}.$$

$$AB = \begin{bmatrix} 0 \\ 1 \\ -7 \end{bmatrix}, \quad A^2B = \begin{bmatrix} 1 \\ -7 \\ 37 \end{bmatrix}.$$

For this system, $n = 3$. Hence, the controllability matrix is given by

$$\mathcal{C}(A, B) = [B \ AB \ A^2B] = \begin{bmatrix} 0 & 0 & 1 \\ 0 & 1 & -7 \\ 1 & -7 & 37 \end{bmatrix}.$$

$\det[\mathcal{C}(A, \ B)] = -1 \neq 0$ and hence $\mathcal{C}(A, \ B)$ is full rank. The system is controllable. Once controllability is guaranteed, next step may be to choose an appropriate control law, $u(t)$.

Example 3.8: A Wheatstone bridge circuit is shown in Fig. 3.5. The bridge can be balanced by varying one of the resistances until the voltmeter reading becomes zero. The condition to be satisfied to balance the bridge is $R_1C_1 = R_2C_2$. Hence, when $R_1C_1 = R_2C_2$, the potential differences across the capacitors are always such that $v_{c_1} = v_{c_2}$.

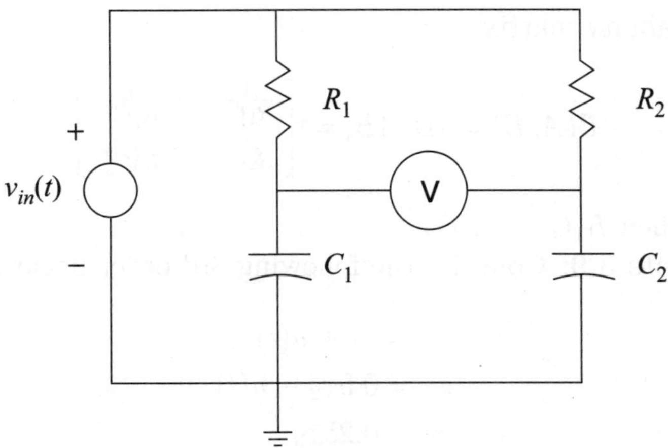

Figure 3.5: Wheatstone bridge circuit.

Hence, if the states for this dynamic system are chosen to be $x_1 = v_{c_1}$, $x_2 = v_{c_2}$, system will be uncontrollable when $R_1C_1 = R_2C_2$. This is because when the bridge is balanced, the states $x_1 = v_{c_1}$ and $x_2 = v_{c_2}$, can only be steered along the line $x_1 = x_2$ in the state space by affecting a control input, $v_{in}(t)$. Since $\boldsymbol{x}(t) = [x_1(t)\ x_2(t)]^T$ can not be steered to any final state, by definition, the system is uncontrollable.

Controllability test for this linear electrical system yields the same result as follows. For the balanced bridge, Kirchoff's current law allows one to write the following equations for the current flow in the circuit

$$C_1 \dot{e}_{c_1} + \frac{v_{c_1} - v_{in}}{R_1} = 0$$

$$C_2 \dot{e}_{c_2} + \frac{v_{c_2} - v_{in}}{R_2} = 0$$

whence rearranged yield

$$\dot{e}_{c_1} = -\frac{1}{R_1 C_1} v_{c_1} + \frac{1}{R_1 C_1} v_{in}$$

$$\dot{e}_{c_2} = -\frac{1}{R_2 C_2} v_{c_2} + \frac{1}{R_2 C_2} v_{in}.$$

When put into state-space form, we get

$$A = \begin{bmatrix} -1/R_1 C_1 & 0 \\ 0 & -1/R_2 C_2 \end{bmatrix} \quad \text{and} \quad B = \begin{bmatrix} 1/R_1 C_1 \\ 1/R_2 C_2 \end{bmatrix}.$$

The controllability matrix

$$\mathcal{C}(A, B) = [B\ AB] = \begin{bmatrix} \frac{1}{R_1 C_1} & -\frac{1}{R_1^2 C_1^2} \\ \frac{1}{R_2 C_2} & -\frac{1}{R_2^2 C_2^2} \end{bmatrix}$$

is singular when $R_1 C_1 = R_2 C_2$.

Example 3.9: Consider the following 3rd order linear system

$$\dot{x}_1 = x_1 + u(t)$$

$$\dot{x}_2 = 0.5 x_2 - u(t)$$

$$\dot{x}_3 = 0.25 x_3$$

for which

$$A = \begin{bmatrix} 1 & 0 & 0 \\ 0 & 0.5 & 0 \\ 0 & 0 & 0.25 \end{bmatrix} \quad \text{and} \quad B = \begin{bmatrix} 1 \\ 1 \\ 0 \end{bmatrix}.$$

It is obvious that no control can be affected on the state x_3 and hence the system is uncontrollable. It is direct to show that $\mathcal{C} = [B\ AB\ A^2B]$ is singular.

Example 3.10: Consider the inverted pendulum control problem detailed in example 3.14 (also see Fig. 3.9). For $M = 1$ kg, $m = 150$ g, and $l = 1$ m, where M, m, and l, respectively, are masses of the cart and the pendulum, and length of the pendulum. The state equation matrices of the linearized model are

$$
A = \begin{bmatrix} 0 & 1 & 0 & 0 \\ 4.45 & 0 & 0 & 0 \\ 0 & 0 & 0 & 1 \\ -0.58 & 0 & 0 & 0 \end{bmatrix} \quad \text{and} \quad B = \begin{bmatrix} 0 \\ -0.40 \\ 0 \\ 0.92 \end{bmatrix}.
$$

The states are $\boldsymbol{x} = [\theta\ \dot{\theta}\ x\ \dot{x}]^T$ with x and θ being the displacement of the cart and the angular position of the pendulum. It is direct to show that $\det[\mathcal{C}(A, B)] \neq 0$ for the controllability matrix

$$
\mathcal{C}(A, B) = [B\ AB\ A^2B\ A^3B] = \begin{bmatrix} 0 & -0.40 & 0 & 1.76 \\ 0.40 & 0 & -1.76 & 0 \\ 0 & 0.92 & 0 & 0.23 \\ 0.92 & 0 & 0.23 & 0 \end{bmatrix}.
$$

Hence, the system is controllable and a control algorithm can be found to move the cart without causing the pendulum to fall. There can be several control solutions to this problem. The option of state feedback control is discussed shortly in example 4.13. Since this controllability result has been obtained using the linearlized model valid for small values of θ, steering the pendulum back to zero may only be possible for small deviations from the pendulum equilibrium position.

Alternative tests for controllability of linear systems include testing whether the A and B are in controllable canonical form (CF) or transformable into CF. Canonical forms will be briefed shortly.

3.4.3 Observability

Once a controllable system is arrived at, the next step is to choose suitable control strategy, $\boldsymbol{u}(t)$, that would give the expected control performance. When selecting the control method or the strategy, one has to take into account the problems in practical implementation such as cost, availability of sensors to measure the variables that need to be fed back, complexity of the control algorithm etc. For instance, state feedback control that will be discussed shortly requires all state variables to be fed back to the controller. The number of state variables can sometimes be excessive making sensing all of them too expensive. Another practical problem can be that not all the state variables are measurable.

Design of an *observer* refers to estimating the state variables from the output variables, $\boldsymbol{y}(t)$, which are measurable. In control literature, the estimated states are often denoted as $\bar{\boldsymbol{x}}(t)$. The condition that such an observer can be designed for the system is called the observability of the system. Essentially, a system will be observable if all the state variables have an influence on some of the outputs. If any one of the states can not be estimated from the measured outputs, the state is said to be unobservable and the system is not completely observable or simply unobservable.

A system is said to be observable, if given any input, $\boldsymbol{u}(t)$, there exists a finite time, $t_f \geq t_0$, such that the knowledge of $\boldsymbol{y}(t)$ and $\boldsymbol{u}(t)$ over $t_0 \leq t < t_f$ are sufficient to determine every state $\boldsymbol{x}(t_0)$. By implication, since

$$\boldsymbol{x}(t) = e^{A(t-t_0)}\boldsymbol{x}(t_0) + \int_{t_0}^{t} e^{A(t-\tau)}Bu(\tau)d\tau \qquad (3.47)$$

once we have $\boldsymbol{x}(t_0)$, we have the state vector for all $t > t_0$.

Consider the linear system given by

$$\dot{x}_1 = -2x_1 + u(t) \qquad (3.48)$$
$$\dot{x}_2 = -x_2 + u(t). \qquad (3.49)$$

Assume that the output equation is $y(t) = x_1(t)$. Once measured $y(t)$ is available, estimating the first state is direct as $x_1(t) = y(t)$. The second state, $x_2(t)$, is not connected to the output, $y(t)$, in any way and hence $x_2(t)$ can not be estimated from the measured information on $y(t)$. The system is unobservable.

Now, let (3.48) be modified as

$$\dot{x}_1(t) = -2x_1 + x_2 + u(t). \qquad (3.50)$$

Since the second state, $x_2(t)$, is now connected to the output, $y(t)$, through the equation (3.50) where $x_1(t) = y(t)$, a suitable observer algorithm may be synthesized. Obviously, $x_2(t_0)$ can be observed using (3.50) when $y(t) = x_1(t)$ and $u(t)$ are known over a period $t_0 \leq t < t_f$.

Condition of observability of a linear system depends on the matrices A and C only.

3.4.4 Testing for Observability

Consider the following linear time-invariant system expressed using familiar notations

$$\dot{x}(t) = Ax(t) + Bu(t) \tag{3.51}$$
$$y(t) = Cx(t). \tag{3.52}$$

For the preceding system to be completely observable or simply observable, it can be shown that the *observability matrix*, $\mathcal{O}(A, C)$, defined as

$$\mathcal{O}(A, C) = \begin{bmatrix} C \\ CA \\ \vdots \\ CA^{n-1} \end{bmatrix} \tag{3.53}$$

must have rank n where $A \in \Re^{n \times n}$. The proof of this result is omitted in this book. Observability of a system only depends on A and C, and is independent of the matrix, B. The observability of the system is sometimes referred to as the observability of the pair $\{A, C\}$.

Example 3.11: Consider the linear system given by (3.48) and (3.49). For this system

$$A = \begin{bmatrix} -2 & 0 \\ 0 & -1 \end{bmatrix}, \quad C = \begin{bmatrix} 1 & 0 \end{bmatrix}, \quad \text{and} \quad CA = \begin{bmatrix} -2 & 0 \end{bmatrix}.$$

Also, $n = 2$. And, the observability matrix given by

$$\mathcal{O}(A, C) = \begin{bmatrix} C \\ CA \end{bmatrix} = \begin{bmatrix} 1 & 0 \\ -2 & 0 \end{bmatrix}$$

is singular. The system is unobservable.

Example 3.12: Consider the linear system discussed a while ago and represented by (3.49) and (3.50)

$$\dot{x}_1 = -2x_1 + x_2 + u(t)$$
$$\dot{x}_2 = -x_2 + u(t)$$
$$y = x_1$$

whence put into state-space form, we get

$$A = \begin{bmatrix} -2 & 1 \\ 0 & -1 \end{bmatrix} \quad \text{and} \quad C = \begin{bmatrix} 1 & 0 \end{bmatrix}.$$

The observability matrix is

$$\mathcal{O}(A, C) = \begin{bmatrix} C \\ CA \end{bmatrix} = \begin{bmatrix} 1 & 0 \\ -2 & 1 \end{bmatrix}$$

and $\det[\mathcal{O}(A, C)] = 1 \neq 0$. System is observable.

Example 3.13: Consider the inverted pendulum control problem detailed in example 3.14 (also see Fig. 3.9). For $M = 1$ kg, $m = 150$ g, and $l = 1$ m, where M, m, and l, respectively, are masses of the cart and the pendulum, and length of the pendulum, the state matrix of the linearized model is

$$A = \begin{bmatrix} 0 & 1 & 0 & 0 \\ 4.45 & 0 & 0 & 0 \\ 0 & 0 & 0 & 1 \\ -0.58 & 0 & 0 & 0 \end{bmatrix}.$$

The states are $\boldsymbol{x} = [\theta \ \dot{\theta} \ x \ \dot{x}]^T$ with x and θ being the displacement of the cart and the angular position of the pendulum. If $y(t) = \theta(t)$, i.e., pendulum angular displacement is the only measurement, then

$$C = [1 \ 0 \ 0 \ 0].$$

It is direct to show that $\det[\mathcal{O}(A, C)] = 0$ and hence the system is unobservable. If measurements of displacement of the cart, $x(t)$, are available, then the state $x_3(t) = x(t)$ will relate to the output, $y(t)$. For simplicity, let us assume that only $x(t)$ is measured. Then

$$C = [0 \ 0 \ 1 \ 0].$$

It is direct to show that $\det[\mathcal{O}(A, C)] \neq 0$. Hence, rest of the states are also observable if at least the measurements of the displacement of the cart, $x(t)$, are available over a finite period of time.

This conclusion can easily be arrived at by having a close look at the equations (3.69) and (3.70). These equations suggest that not only the measurements but also the control input, $u(t)$, should be known over the finite period of time to observe the nonmeasurable states.

Estimated values of the states by the use of an observer may not exactly be the same as the actual values. One key reason is that only linearized or simplified models of the systems are often used to design the observers. For instance, the aforementioned observability analysis is based on a linearized model of the plant. Observers designed using the linearized models will only be approximate. For these reasons, in control literature, the estimated value of $(\cdot)$ is normally denoted by $(\hat{\cdot})$. In order to design better observers, the designer may take into account the non-simplified original model. For instance, in the inverted pendulum case, the original model consists of trigonometric terms. Implementing trigonometric terms is difficult using analog devices, but rather straightforward in the digital implementation. One of the key advantages of digital control is that complex algorithms can be implemented with relative ease.

Alternative tests for observability of linear systems includes testing whether the A and C are in observable canonical form (OF) or transformable into OF.

3.5 State Feedback Control

Once a controllable system is arrived at, the next step is to control the system in some way or other. Among numerous control methods, one may resort to PID control or its variants, fuzzy-logic control, or pole placement method. In pole-placement method, the desired control performance is achieved by placing the closed-loop poles at desirable locations of the s-plane.

In the state feedback control of a linear plant, $\dot{\boldsymbol{x}}(t) = A\boldsymbol{x}(t) + Bu(t)$, shown in Fig. 3.6, state variables are fed back to the input side through a feedback gain matrix, $K = [k_1 \ k_2 \cdots k_n]$. With familiar notations, $r(t)$ is the reference input.

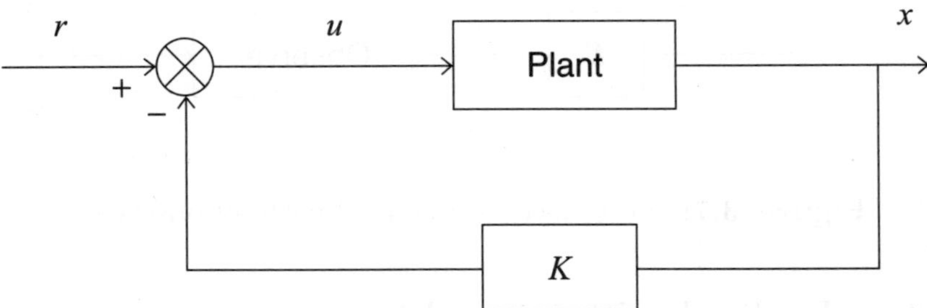

Figure 3.6: State feedback control.

Equation of the closed-loop system can be obtained as follows. The control, $u(t)$, is given by

$$u(t) = -K\boldsymbol{x}(t) + r(t) \tag{3.54}$$

whence substituted in the the plant equation yields

$$\dot{\boldsymbol{x}}(t) = A\boldsymbol{x}(t) + B[-K\boldsymbol{x}(t) + r(t)]. \tag{3.55}$$

This is rearranged to obtained the closed-loop system equation as

$$\dot{\boldsymbol{x}}(t) = (A - BK)\boldsymbol{x}(t) + Br(t). \tag{3.56}$$

The design approach in this case is to determine the feedback gain, K, such that the desired performance characteristics are achieved by placing the closed-loop poles or eigenvalues of $A - BK$ at suitable locations on the s-plane.

   This method requires all the state variables to be available, which can well be a strong requirement in practice due to cost and unavailability of sensors. Therefore, it may sometimes be necessary to design an observer to estimate the state vector from the output vector. Block diagram representation of a control system with state-feedback and observer is shown in Fig.3.7 where $\bar{\boldsymbol{x}}$ are the estimated states. Observers can be designed to estimate the states that are observable.

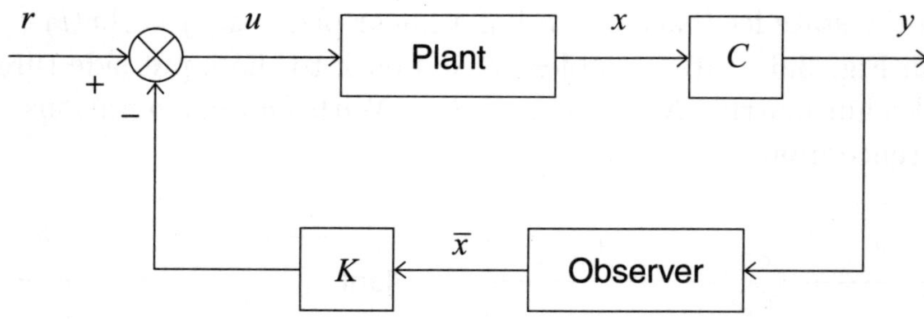

Figure 3.7: State feedback control with an observer.

3.5.1 State Feedback with Integral Control

One major limitation of conventional state feedback control can be the large steady-state error. Therefore, state feedback control is only suitable for certain regulator systems. One solution to this problem is to introduce an integral control action, just as with PI controller. In the method called state feedback with integral control, an integral control action is added by inserting an integrator that integrates the error signal.

   Block diagram representation of state feedback with integral control is given in Fig. 3.8. $K = [k_1 \ k_2 \cdots k_n]$ are state feedback gains and k_{n+1}

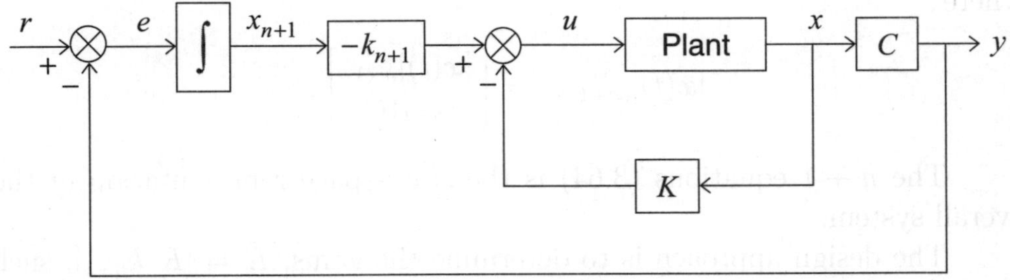

Figure 3.8: State feedback with integral control.

is the scalar integral gain. Consider a single input single output (SISO) system, for simplicity. The following equations can be written.

$$\dot{\boldsymbol{x}}(t) = A\boldsymbol{x}(t) + Bu(t) \tag{3.57}$$

$$\dot{x}_{n+1}(t) = r(t) - y(t) \tag{3.58}$$

$$y(t) = C\boldsymbol{x}(t) \tag{3.59}$$

$$u(t) = -K\boldsymbol{x}(t) - k_{n+1}x_{n+1}(t). \tag{3.60}$$

Substituting (3.60) into (3.57) yields

$$\dot{\boldsymbol{x}}(t) = A\boldsymbol{x}(t) + B[-K\boldsymbol{x}(t) - k_{n+1}x_{n+1}(t)]$$
$$= [A - BK]\boldsymbol{x}(t) - Bk_{n+1}x_{n+1}(t). \tag{3.61}$$

From (3.58) and (3.59), we get

$$\dot{x}_{n+1}(t) = C\boldsymbol{x}(t) + r(t). \tag{3.62}$$

By stacking (3.61) and (3.62) together, we get

$$\begin{bmatrix} [\dot{\boldsymbol{x}}]_{n\times 1} \\ \dot{x}_{n+1} \end{bmatrix} = \begin{bmatrix} A - BK & -Bk_{n+1} \\ -C & 0 \end{bmatrix} \begin{bmatrix} [\boldsymbol{x}]_{n\times 1} \\ x_{n+1} \end{bmatrix} + \begin{bmatrix} 0_{n\times 1} \\ 1 \end{bmatrix} r(t). \tag{3.63}$$

Noting that

$$\begin{bmatrix} A - BK & -Bk_{n+1} \\ -C & 0 \end{bmatrix} = \begin{bmatrix} A & 0 \\ -C & 0 \end{bmatrix} - \begin{bmatrix} B \\ 0 \end{bmatrix} \begin{bmatrix} K & k_{n+1} \end{bmatrix}$$
$$= \bar{A} - \bar{B}\bar{K}$$

(3.63) can also be written as

$$\dot{\bar{\boldsymbol{x}}}(t) = (\bar{A} - \bar{B}\bar{K})\bar{\boldsymbol{x}}(t) + \begin{bmatrix} 0_{n\times 1} \\ 1 \end{bmatrix} r(t) \tag{3.64}$$

where

$$[\bar{\boldsymbol{x}}(t)]_{(n+1)\times 1} = \begin{bmatrix} [\boldsymbol{x}(t)]_{n\times 1} \\ x_{n+1}(t) \end{bmatrix}.$$

The $n+1$ equations (3.64) is the state-space representation of the overall system.

The design approach is to determine the gains, $\bar{K} = [K \; k_{n+1}]$, such that the desired performance characteristics are achieved by placing the eigenvalues of $\bar{A} - \bar{B}\bar{K}$ at suitable locations on the s-plane.

Example 3.14: Inverted pendulum is widely used as a laboratory experimental setup to carry out control experiments. It is used in simulations and experiments to show the performance of different controllers such as PID, state feedback, and fuzzy controllers etc. The complete system is shown in Fig. 3.9. An inverted pendulum is a cylindrical bar with its pivot mounted on a cart so that it is free to rotate around the pivot in the vertical plane of the cart's moving direction. The cart is driven by an electric motor. The pendulum is unstable if the cart is stationary but can be kept upright by applying a proper control force, $u(t)$, which is the external force on the cart due to the interaction of the drive wheels with the contact surface. Hence, $u(t) = \tau(t)/R$ with $\tau(t)$ and R being the motor torque on the wheels and the wheel radius, respectively.

The total system consisting of the inverted pendulum and the cart can be mathematically modeled if the system parameters are known. Let the system parameters be

M = mass of the cart

m = mass of the pendulum

$2l$ = length of the pendulum

I = moment of inertia of the pendulum about its center of gravity.

Assume that the cart's center-of-gravity and the pivot are located on the same vertical axis. Let the horizontal displacement of this line be $x(t)$. The dynamic model can be obtained using the following equations, (3.65) through (3.68).

Writing Euler equation for the rotational motion of the pendulum about its center-of-gravity yields

$$I\frac{d^2\theta}{dt^2} = V(t)l\sin\theta(t) - H(t)l\cos\theta(t) \tag{3.65}$$

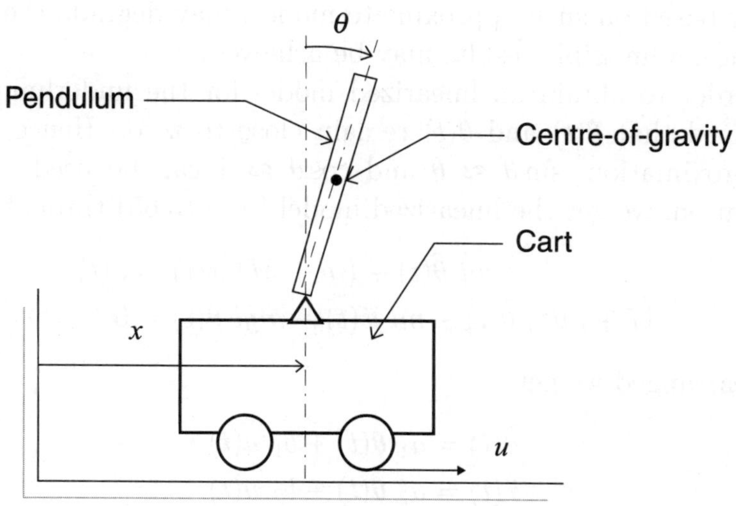

Figure 3.9: Inverted pendulum.

where $V(t)$ and $H(t)$ are the vertical and horizontal reaction forces at the pivot. Writing Newton equations for the linear motion of the pendulum, respectively, in horizontal and vertical directions yield

$$m\frac{d^2}{dt^2}(x(t) + l\sin\theta(t)) = H(t) \qquad (3.66)$$

$$m\frac{d^2}{dt^2}(l\cos\theta(t)) = V(t) - mg \qquad (3.67)$$

where $g = 9.81\text{ms}^{-2}$ is the acceleration of gravity. Equation of linear motion of the cart is

$$M\frac{d^2x(t)}{dt^2} = u(t) - H(t). \qquad (3.68)$$

By removing internal forces $H(t)$ and $V(t)$ in (3.65) through (3.68), one can obtain the dynamic model of the cart-pendulum system.

An approximate model of the controlled system is often considered in the controller design. Such approximations are often achieved by linearizing nonlinear equations. One way to obtain a linear approximate model is by expanding the original nonlinear model into a Taylor series about the normal operating point and dropping out the higher order terms. When the nonlinearities are due to trigonometric terms, the equations are linearized by using small angle approximation. Even though design of controllers based on such approximate models may degrade the controller performance, admissible results may be achieved.

In order to obtain an linearized model for the underlining system, it is assumed that $\theta(t)$ and $\dot{\theta}(t)$ remain close to zero. Hence, the small angle approximation, $\sin\theta \approx \theta$ and $\cos\theta \approx 1$ can be used. With this approximation, we get the linearized model from (3.65) through (3.68) as

$$ml\,\ddot{\theta}(t) + (m + M)\,\ddot{x}(t) = u(t)$$
$$(I + ml^2)\,\ddot{\theta}(t) + ml\,\ddot{x}(t) - mgl\,\theta(t) = 0$$

whence rearranged we get

$$\ddot{\theta}(t) = a_1\,\theta(t) + b_1\,u(t) \tag{3.69}$$
$$\ddot{x}(t) = a_2\,\theta(t) + b_2\,u(t) \tag{3.70}$$

where the coefficients a_1, a_2, b_1, and b_2 can be calculated if the system parameters M, m, l, and I are known. It is obvious from (3.69) and (3.70) also that both the cart position, $x(t)$, and the pendulum angular position, $\theta(t)$, are governed by a single control input, $u(t)$. Further, the two dynamics are coupled. We normally control the pendulum position upright by applying a suitable control input, $u(t)$. Due to the coupled nature of dynamics, the cart position will vary in an uncontrolled way.

Choose the states as $x_1(t) = \theta(t)$, $x_2(t) = \dot{\theta}(t)$, $x_3(t) = x(t)$, and $x_4(t) = \dot{x}(t)$, we obtain the following state-space representation of the linearized system

$$\dot{\boldsymbol{x}}(t) = A\boldsymbol{x}(t) + Bu(t) \tag{3.71}$$

where

$$A = \begin{bmatrix} 0 & 1 & 0 & 0 \\ a_1 & 0 & 0 & 0 \\ 0 & 0 & 0 & 1 \\ a_2 & 0 & 0 & 0 \end{bmatrix}, \quad B = \begin{bmatrix} 0 \\ b_1 \\ 0 \\ b_2 \end{bmatrix}.$$

Assume that $M = 1$ kg, $m = 150$ g, $l = 1$ m. Inertia can be obtained using $I = \frac{1}{3}m(2l)^2$. For these values, $a_1 = 4.45$, $a_2 = -0.58$, $b_1 = -0.40$, and $b_2 = 0.92$. Computing the eigenvalues of A, it is direct to show that the system without feedback control is unstable. This is well-known common sense. In order to move the cart from one location to another without causing pendulum to fall, various control algorithms can be used to realize a suitable control action, $u(t)$. Let the control be realized by state feedback as

$$u(t) = -K\boldsymbol{x}(t) \tag{3.72}$$

where $K = [k_1 \ k_2 \ k_3 \ k_4]$ is the constant gain matrix. The characteristic equation of the closed-loop system with state feedback is

$$|sI - A + BK| = \begin{vmatrix} 0 & -1 & 0 & 0 \\ -a_1 + b_1 k_1 & s + b_1 k_1 & b_1 k_3 & b_1 k_4 \\ 0 & 0 & s & -1 \\ -a_2 + b_2 k_1 & b_2 k_2 & b_2 k_3 & s - b_2 k_4 \end{vmatrix} = 0 \quad (3.73)$$

which is a 4th order polynomial of s.

Design objective in this case is to find the feedback gain matrix, K, such that the closed-loop system is stable. Since the system model is available, one can perform computer simulations to obtain K. After few trial-and-error runs using the function *svdesign* of MATLAB, one can find the roots of the characteristic equation that satisfy the design requirements. Let those roots be s_1, s_2, s_3, and s_4. The corresponding characteristic equation is

$$(s - s_1)(s - s_2)(s - s_3)(s - s_4) = 0. \quad (3.74)$$

Equating like coefficients of the equations (3.73) and (3.74), and solving the resulting simultaneous equations, one can get k_1, k_2, k_3, and k_4 of the feedback gain matrix. These values can be further tuned doing experimental testing on the real world system. This may be essential as the dynamic models are not accurate.

Since the controller design is based on a linearized model obtained using the small angle approximation, such a controller may not be able to stabilize the system, for instance, when the initial conditions $\theta(0)$ and/or $\dot{\theta}(0)$ are large.

Implementation of the state-feedback control requires all the state variable to be available. In laboratory setups, the cart position and the pendulum angle may be obtained through potentiometers. Velocities may be approximated using numerical methods such as the finite difference method.

3.6 Canonical Forms

Controllability and observability are actually properties of a system representation and not of a system per se. A system representation may be transformed into one that is controllable/observable. These forms are called canonical form. We look at two such important canonical forms, namely, the controllability form (CF) and observability form (OF).

For SISO systems, the controllability canonical form can be obtained by reading coefficients from the system transfer function. If the A, B, C

matrices of the state-space representation of a SISO system is available, one can easily obtain the transfer function (see equation (3.75)) by taking the Laplace transform of the state and output equations and assuming zero initial conditions. Alternatively, the Matlab command SS2TF can be used. If the system transfer function takes the following form

$$\frac{Y(s)}{U(s)} = C(sI - A)^{-1}B = \frac{b_{n-1}s^{n-1} + \cdots + b_1 s + b_0}{s^n + a_{n-1}s^{n-1} + \cdots + a_s + a_0} \tag{3.75}$$

the system representation in the CF is

$$\dot{\boldsymbol{x}}_c(t) = A_c \boldsymbol{x}_c(t) + B_c u(t) \tag{3.76}$$

$$y(t) = C_c \boldsymbol{x}(t) \tag{3.77}$$

where

$$A_c = \begin{bmatrix} 0 & 1 & 0 & \cdots & 0 \\ 0 & 0 & 1 & \cdots & 0 \\ \vdots & \vdots & \vdots & \vdots & \vdots \\ 0 & 0 & 0 & \cdots & 1 \\ -a_0 & -a_1 & -a_2 & \cdots & -a_{n-1} \end{bmatrix}, \quad B_c = \begin{bmatrix} 0 \\ 0 \\ \vdots \\ 0 \\ 1 \end{bmatrix}, \quad C_c = \begin{bmatrix} b_0 \\ b_1 \\ b_2 \\ \vdots \\ b_{n-1} \end{bmatrix}^T.$$

The observability canonical form (OF) is

$$\dot{\boldsymbol{x}}_o(t) = A_o \boldsymbol{x}_o(t) + B_o u(t) \tag{3.78}$$

$$y(t) = C_o \boldsymbol{x}(t) \tag{3.79}$$

where

$$A_o = A_c^T$$
$$B_o = C_c^T$$
$$C_o = B_c^T.$$

The state variables, $\boldsymbol{x}(t)$, in the original state-space model that is usually derived through mathematical modeling of the system are real physical variables internal to the system. However, $\boldsymbol{x}_c(t)$ and $\boldsymbol{x}_o(t)$ above may not be real physical internal variables of the system. Since the above representations are derived from the transfer function, the internal physical structure of the system is generally not known anyway. Nevertheless, CF and OF are sometimes useful and exhibit some useful mathematical properties. If the internal structure of the system is known, the state-space representation derived using first principles in the real physical variables might be the preferred option.

Methods do exist to transform the system representation directly from the original state-space representation into canonical forms without going through the transfer function. Such methods are hence general and can be used for MIMO systems as well. One such method is as follows. Write the characteristics equation of the system

$$|sI - A| = s^n + a_{n-1}s^{n-1} + \cdots + a_s + a_0 \tag{3.80}$$

and matrices A_c and B_c are the same as in (3.78). Now, we have the controllability matrix of the CF given by

$$\mathcal{C}_c(A_c, B_c) = [B_c \ A_cB_c \ \ldots A_c^{n-1}B_c]. \tag{3.81}$$

C_c of the CF remains to be found. The mapping rule may be written as

$$\boldsymbol{x} = T\boldsymbol{x}_c$$

where T is the transformation matrix. Hence, the relationship between original state representation and the CF is

$$A_c = T^{-1}AT, \quad B_c = T^{-1}B, \text{ and } C_c = CT.$$

If T is known, C_c can be obtained. Rewrite $\mathcal{C}_c(A_c, B_c)$ as

$$\begin{aligned}
\mathcal{C}_c(A_c, B_c) &= [B_c \ A_cB_c \ \ldots A_c^{n-1}B_c] \\
&= [T^{-1}B \ T^{-1}ATT^{-1}B \ \ldots] \\
&= T^{-1}[B \ AB \ \ldots] = T^{-1}\mathcal{C}(A, B)
\end{aligned}$$

where $\mathcal{C}(A, B)$ is the controllability matrix of the original state-space representation. To this end, we have

$$\mathcal{C}_c = T^{-1}\mathcal{C} \text{ and } T = \mathcal{C}\mathcal{C}_c^{-1}.$$

Hence, $C_c = CT$ can be obtained avoiding transfer function calculation.

The CF has the advantage of being controllable and the OF has the advantage of being observable. Most of the well-formed systems, however, are both controllable and observable. The feature of controllability was discussed previously in this chapter. Recall how the feature of observability was important, for instance, in the design of state-feedback control.

Chapter 4

Digital Control Theory

4.1 Background

In digital computer based controller implementation, the signals are sampled in time and quantized in magnitude. Sampling makes the signals noncontinuous and hence the differential equations that describe system behavior becomes difference equations. Even though the difference between differential equations and difference equations can be minimized by sufficiently increasing the sampling frequency, this may not always be possible due to practical reasons. To that end, the effects of finite sampling on the control theory need to be addressed. This gives rise to the sub-discipline of digital control theory.

4.2 Mathematical Methods of Discrete Systems

4.2.1 A Sampled Signal

Unit impulse function is important to mathematically represent a sampled signal. Unit impulse function is defined as

$$\delta(k) = \begin{cases} 1 & \text{if } k = 0 \\ 0 & \text{otherwise} \end{cases}$$

and shown in Fig. 4.1(a). Here, k is the time index. A sampler, a continuous signal, and its sampled signal are shown in Fig. 4.1.

According to Fig. 4.1, a sampled signal can be written as

$$x(kT) = x(0)\delta(kT) + x(T)\delta(k-1)T + x(2T)\delta(k-2)T + \cdots \to \infty. \quad (4.1)$$

If the samples are taken at regular intervals, the sampling period T, is often omitted in the equation for brevity. Hence

$$x(k) = x(0)\delta(k) + x(1)\delta(k-1) + x(2)\delta(k-2) + \cdots \to \infty \quad (4.2)$$

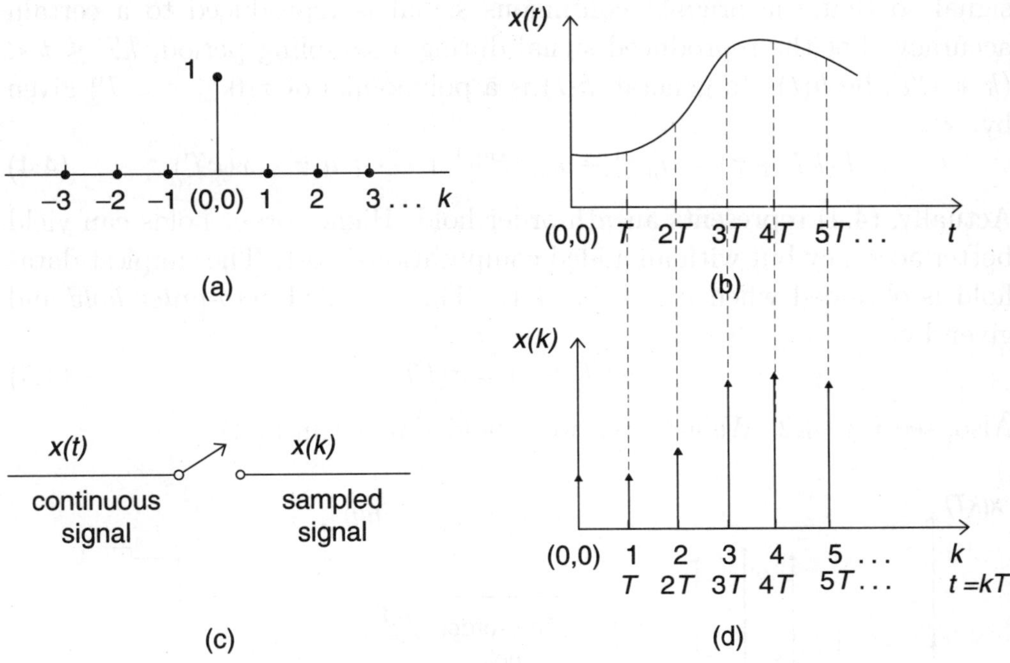

Figure 4.1: (a) Unit pulse function, (b) a continuous signal, (c) a sampler, and (d) the sampled signal.

or

$$x(k) = \sum_{n=0}^{\infty} x(n)\, \delta(k-n) \tag{4.3}$$

which is a finite series of shifted unit pulses each weighted with a term $x(n)$.

4.2.2 Sampling and Data Hold

Sample-and-hold circuit samples an analog signal and holds this signal (at a constant value by a zero-order hold) for a single sampling period. A sampler-and-hold circuit is an integral part of most analog-to-digital converters(ADCs).

In a sampler, a switch closes at regular intervals to admit the value of the signal. A sampler converts a continuous signal into a train of pulses separated by the sampling period, T. Sampler transmits no information between two consecutive sampling instants. In practice, sampling period is very short and the variations of the continuous signal in between two consecutive sampling instants may be ignored. Otherwise, a hold circuit or an extrapolator can be used to convert the sampled signal into a continuous signal so that the original continuous signal is reproduced to a certain accuracy. Let the reproduced signal during a sampling period, $kT \leq t < (k+1)T$, be $h(t)$. In general, $h(t)$ is a polynomial of $\tau(0 \leq \tau < T)$ given by

$$h(kT + \tau) = a_n \tau^n + a_{n-1} \tau^{n-1} + \cdots + a_1 \tau + x(kT). \tag{4.4}$$

Actually, (4.4) represents an nth-order hold. Higher order holds can yield better accuracy but with an added computational cost. The simplest data-hold is obtained when $n = 0$ in (4.4). This is called *zero-order hold* and given by

$$h(kT + \tau) = x(kT). \tag{4.5}$$

Also, see Fig. 4.2. We get first-order hold if $n = 1$ in (4.4).

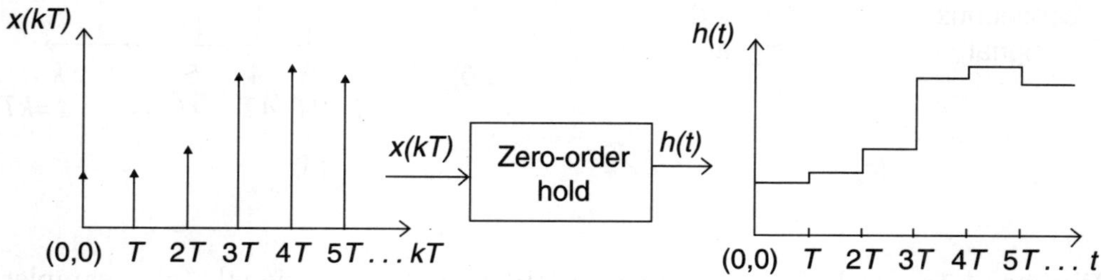

Figure 4.2: Zero-order hold.

The transfer function of the zero-order hold, $G_{zoh}(s) = \frac{H(s)}{X^*(s)}$, is important to analyze systems with a hold circuit after the sampler. For the zero-order hold, the output can be written as

$$h(t) = x(0)[u(kT) - u(k-1)T] + x(T)[u(k-1)T - u(k-2)T]$$
$$+ x(2T)[u(k-3)T - u(k-3)T] + \cdots$$

Since

$$\mathcal{L}[u(k-n)T] = \frac{e^{-nTs}}{s}$$

the Laplace transform of the preceding equation becomes

$$\mathcal{L}[h(t)] = H(s) = \sum_{n=0}^{\infty} x(nT) \frac{e^{-nTs} - e^{-(n+1)Ts}}{s}$$
$$= \frac{1 - e^{-Ts}}{s} \sum_{n=0}^{\infty} x(nT) e^{-nTs}. \tag{4.6}$$

In (4.9), it will be shown that the second half of (4.6) is the $\mathcal{L}[x(nT)] = X^*(s)$. Hence, (4.6) can be rewritten as

$$H(s) = \frac{1 - e^{-Ts}}{s} X^*(s)$$

suggesting that the transfer function, $G_{zoh}(s)$, of zero-order hold is

$$\frac{H(s)}{X^*(s)} = G_{zoh}(s) = \frac{1 - e^{-Ts}}{s}. \tag{4.7}$$

4.2.3 The z-transform

The z-tansform is the discrete time counterpart of the well-known Laplace transform. An in-depth discussion of Laplace transform is omitted in this book as it is an essential integral part of an engineering mathematics syllabus in the undergraduate level. In this section, I assume that the reader has the basic background of Laplace transform, partial fractions, using the Laplace tables, and properties of Laplace transform.

Consider a causal signal defined for $t \geq 0$. Since $t = kT$; $k = 1, 2, 3 \cdots \infty$, the sampled signal in (4.1) may also be written as

$$x^*(t) = x(0)\delta(t) + x(T)\delta(t - T) + x(2T)\delta(t - 2T) + \cdots \rightarrow \infty \tag{4.8}$$

where $\delta(t)$ is an unit impulse at $t = 0$ and $\delta(t - nT)$ is an unit impulse at $t = nT$.

Let $x^*(t) \leftrightarrow X^*(s)$ be a Laplace transform pair. The Laplace transform of the sampled signal (4.8) is

$$X^*(s) = x(0) + x(T)e^{-Ts} + x(2T)e^{-2Ts} + + \cdots \rightarrow \infty$$
$$= \sum_{n=0}^{\infty} x(nT)\, e^{-nTs}. \tag{4.9}$$

Define

$$z = e^{Ts} \quad \text{where } s = \sigma + j\omega \text{ is the Laplace operator.}$$

Now, the preceding equation can be rewritten as

$$X(z) = \sum_{n=0}^{\infty} x(nT)\, z^{-n}. \tag{4.10}$$

This is the z-transform of the sampled signal $x(kT)$. For example, the z-transform of the unit step function defined as

$$u(k) = \begin{cases} 1 & \text{if } k \geq 0 \\ 0 & \text{otherwise} \end{cases}$$

can be obtained as

$$U(z) = \sum_{n=0}^{\infty} u(k)\, z^{-k}$$

$$= \sum_{n=0}^{\infty} z^{-k} = \frac{z}{z-1}.$$

The mathematical methods used to analyze discrete and continuous systems are quite similar. The z-transform of some functions common in control theory are given in Table 4.1. The multiplier $u(k)$ is used to imply that the results in Table 4.1 are valid only for causal signals.

Table 4.1: z-transform of some common functions

Function, $x(kT)$	z-transform, $X(z)$
$\delta(kT)$	1
$u(kT)$	$\frac{z}{z-1}$
$a^{kT}u(kT)$	$\frac{z}{z-a}$
$e^{-akT}u(kT)$	$\frac{z}{z-e^{-aT}}$
$\sin k\omega T\ u(kT)$	$\frac{z\sin\omega T}{z^2-2z\cos\omega T+1}$
$\cos k\omega T\ u(kT)$	$\frac{z(z-\cos\omega T)}{z^2-2z\cos\omega T+1}$

4.2.4 Properties of the z-transform

Let $\mathcal{Z}\{x(k)\} = X(z) \leftrightarrow x(k)$ be a z-transform pair. The multiplication theorem

$$\mathcal{Z}\{cx(k)\} = c\mathcal{Z}\{x(k)\} \tag{4.11}$$

where c is a constant is obvious from the definition of z-transform as

$$\mathcal{Z}\{cx(k)\} = \sum_{n=0}^{\infty} cx(k)\, z^{-k}$$

$$= c\sum_{n=0}^{\infty} x(k)\, z^{-k} = c\mathcal{Z}\{x(k)\}.$$

Likewise, it is direct to prove the linearity theorem

$$\mathcal{Z}\{c_1 x(k) + c_2 y(k)\} = c_1 X(z) + c_2 Y(z) \tag{4.12}$$

where c_1, c_2 are constants.

The convolution of two discrete causal signals is defined as

$$x(k) * y(k) = \sum_{i=0}^{k} x(i)y(k - i).$$

The convolution theorem states that

$$\mathcal{Z}\{x(k) * y(k)\} = X(z)Y(z). \tag{4.13}$$

Consider

$$X(z) = \sum_{n=0}^{\infty} x(k)\, z^{-k}$$
$$= x(0) + x(1)z^{-1} + x(2)z^{-2} + \cdots \to \infty.$$

Hence

$$\lim_{z \to \infty} X(z) = x(0). \tag{4.14}$$

This is called the initial value theorem.

The final value theorem states that

$$\lim_{z \to 1}(z - 1)X(z) = \lim_{k \to \infty} x(k). \tag{4.15}$$

The time-shift theory can be used to find the z-transform of a discrete function that is shifted $n(> 0)$ time units to the right.

$$x(k - n) \leftrightarrow z^{-n}X(z). \tag{4.16}$$

4.2.5 Inverse z-transform

The process of obtaining the discrete time signal when its z-transform is available is called the inverse z-transform. As the following example shows, the procedure is quite analogous to that of inverse Laplace transform. We use partial fraction, z-transform tables, and the properties just mentioned. For example, consider the following z-transform

$$X(z) = \frac{z}{z^2 - 11z + 10} = \frac{z}{(z - 1)(z - 10)}$$
$$= \frac{-1/9}{z - 1} + \frac{10/9}{z - 10}$$
$$= -\frac{1}{9}\frac{z}{z - 1}z^{-1} + \frac{10}{9}\frac{z}{z - 10}z^{-1}.$$

Using table 4.1 and the time-shift property, the discrete time signal is obtained as

$$x(k) = -\frac{1}{9}u(k-1) + \frac{10}{9}10^{k-1}u(k-1). \qquad (4.17)$$

In z-transform, it is much convenient to obtain partial fraction expansion of $X(z)/z$ and proceed. This way one can avoid the need to apply time-shift property as follows

$$\frac{X(z)}{z} = \frac{1}{(z-1)(z-10)}$$
$$= \frac{-1/9}{z-1} + \frac{1/9}{z-10}$$

and hence

$$X(z) = -\frac{1}{9}\frac{z}{z-1} + \frac{1}{9}\frac{z}{z-10}.$$

Using table 4.1, we get

$$x(k) = -\frac{1}{9}(1-10^k)u(k). \qquad (4.18)$$

Since $x(0) = 0$, (4.17) and (4.18) give the same result.

Let us see how Matlab can help solve difficult inverse z-transform problems. Even though Matlab does not have a command to directly do inverse z-transformations, thorny partial fractions expansion part can be solved as demonstrated in the following example. Let the z-transform be

$$X(z) = \frac{z}{z^2 - 11z + 10} = \frac{z^{-1}}{1 - 11z^{-1} + 10z^{-2}}.$$

The following code in the Matlab command window

```
>> A = [1 − 11 10];
>> B = [0 1];
>> [N, D, K] = residuez(B, A)
```

returns

$$N =$$
$$0.1111$$

$$D = \begin{array}{c} -0.1111 \\ \\ 10 \end{array}$$

$$K = \begin{array}{c} 1 \\ \\ 0 \end{array}$$

where K is the term to be added to $X(z)$ if the order of numerator is greater than that of the denominator. The vectors, N and D are used as follows.

$$
\begin{aligned}
X(z) &= \frac{0.1111}{1 - 10z^{-1}} + \frac{-0.1111}{1 - z^{-1}} \\
&= \frac{0.1111z}{z - 10} - \frac{0.1111z}{z - 1} \\
&= -\frac{1}{9}\frac{z}{z - 1} + \frac{1}{9}\frac{z}{z - 10}
\end{aligned}
$$

which will give the same result as (4.18).

4.2.6 Obtaining the z-transform when the Function in s is Known

It is not always realistic to translate a function in s, $X(s)$, to the corresponding function in z, $X(z)$, directly using $z = e^{Ts}$ or $s = \frac{1}{T}\ln z$ because this will transform finite polynomials to infinite polynomials.

Approximate relationships such as bilinear transformation (2.32) can be used. Another approach is to expand $X(s)$ into partial fractions and use z-transform table to find the z-transform of each term. In other words, convert $X(s)$ into $x(t)$ and then $x(kT)$ into $X(z)$.

Example 4.1: Consider

$$X(s) = \frac{1}{s(s + 1)}.$$

Taking the inverse Laplace transform yields

$$x(t) = 1 - e^{-t} \quad t \geq 0.$$

Hence, substituting $t = kT$ where T is the sampling period, we can obtain the z-transform as

$$
\begin{aligned}
X(z) = \mathcal{Z}[1 - e^{-kT}] &= \frac{1}{1 - z^{-1}} - \frac{1}{1 - e^{-T}z^{-1}} \\
&= \frac{(1 - e^{-T})z}{(z - 1)(z - e^{-T})}.
\end{aligned}
\tag{4.19}
$$

It is noteworthy that for a time varying discrete signal, the z-transform is a function of the sampling period, T. Generally, smaller the T value or higher the sampling frequency, $f_s = 1/T$, the closer the discrete system approximates the original continuous system.

Matlab has a command to convert a continuous system to its discrete from. The syntax is

$$[\text{numd, dend}] = \text{c2dm(num, den, T, 'method')}.$$

For instance, to obtain $X(z)$ when $X(s)$ is known, rewrite $X(s)$ in the following form.

$$X(s) = \frac{1}{s(s+1)} = \frac{1}{s^2 + s}.$$

Assume $T = 1$ s and the method is impulse sampler. The following code in the Matlab command window

```
>> num = [1];
>> den = [1 1 0];
>> T = 1;
>> [numd, dend] = c2dm(num, den, T,'imp')
```

returns

$$\text{numd} = 0 \qquad 0.6321 \quad 0$$
$$\text{dend} = 1.0000 \quad -1.3679 \quad 0.3679.$$

Therefore,

$$X(z) = \frac{0.6321z}{z^2 - 1.3679z + 0.3679}.$$

This is the same as (4.19) evaluated for the sampling period of 1 second, $T = 1$.

Example 4.2: A more typical situation would be when a transfer function, say $X(s)$, is preceded by a sampler and a zero-order hold as follows.

$$G(s) = G_{zoh}X(s) = \frac{1 - e^{Ts}}{s}\frac{1}{s(s+1)}.$$

The z-transform can be obtained as follows.

$$G(z) = (1 - z^{-1})\mathcal{Z}\left(\frac{1}{s^2(s+1)}\right)$$

and to get the z-transform of the term in the braces, one may first invert the Laplace transform to get its continuous time function, say $f(t)$. Then, this time function may be discretized to take $f(kT)$ as follows.

$$\frac{1}{s^2(s+1)} = \frac{A}{s} + \frac{B}{s^2} + \frac{C}{s+1} = -\frac{1}{s} + \frac{1}{s^2} + \frac{1}{s+1}.$$

Thus

$$f(t) = -u(t) + tu(t) + e^{-t}u(t)$$

and

$$f(kT) = -u(kT) + kTu(kT) + e^{-kT}u(kT).$$

Hence, the z-transform $G(z)$ is

$$G(z) = (1 - z^{-1})\left(-\frac{z}{z+1} + \frac{Tz}{(z+1)^2} + \frac{z}{(z - e^{-T})}\right). \quad (4.20)$$

Assume $T = 0.1$ s and the method is impulse sampler followed by zero-order hold. The following code in the Matlab command window

```
>> num = [1];
>> den = [1 1 0];
>> T = 0.1;
>> [numd, dend] = c2dm(num, den, T,'zoh')
```

returns

$$numd = 0 \qquad 0.0048 \quad 0.0047$$
$$dend = 1.0000 \quad -1.9048 \quad 0.9048.$$

Therefore,

$$G(z) = \frac{0.0048z + 0.0047}{z^2 - 1.9048z + 0.9048}.$$

This is the same as (4.20) evaluated for the sampling period of 0.1 second, $T = 0.1$.

4.3 Discrete Time Transfer Function

The definitions of transfer function, characteristics equation, analysis based on locations of poles and zeros for the discrete time system are quite similar and quite analogous to those in the continuous time systems. Various methods were discussed in the preceding chapters to analyze and design continuous time systems. Since we are already familiar with Laplace domain analysis of linear continuous time systems, methods to analyze discrete time systems in the z-domain can now be deducted as methods to map s-domain to z-domain are known.

Quite similar to continuous time systems, the transfer function of a discrete time system is defined as

$$M(z) = \frac{Y(z)}{R(z)} \tag{4.21}$$

where $Y(z)$ and $R(z)$, respectively, are the z-transforms of the output of the system and of the input to the system. $M(z)$ also is a rational function of z and hence can be decomposed into partial fractions and can be used to analyze the system stability and response characteristics in a way quite analogous to the analysis of continuous time systems. Further, the system response to various inputs can be found. The locations of poles and zeros of $M(z)$ govern the system response characteristics.

For example to find the impulse response, set $R(z) = 1$. The output can now be obtained by inverting the z-transform as $Y(z) = M(z) \leftrightarrow y(kT)$ where $y(kT)$ is the impulse response.

For a step input,

$$R(z) = \frac{z}{z - 1}$$

and the step response, $y(kT)$, can be obtained by inverse z-transforming

$$Y(z) = \frac{z}{z - 1} M(z)$$

if the transfer function, $M(z)$, is known.

Example 4.3: Matlab commands exist to obtain system responses to typical inputs if the system transfer function is known. Consider the following discrete transfer function of a system

$$M(z) = \frac{0.0048z + 0.0047}{z^2 - 1.9z + 0.9095}.$$

The following code in the Matlab command window

$$\gg \text{numd} = [0 \quad 0.0048 \quad 0.0047];$$
$$\gg \text{dend} = [1 \quad -1.9 \quad 0.9095];$$
$$\gg \text{dstep(numd, dend)}$$

yields the step response shown in Fig. (4.3).

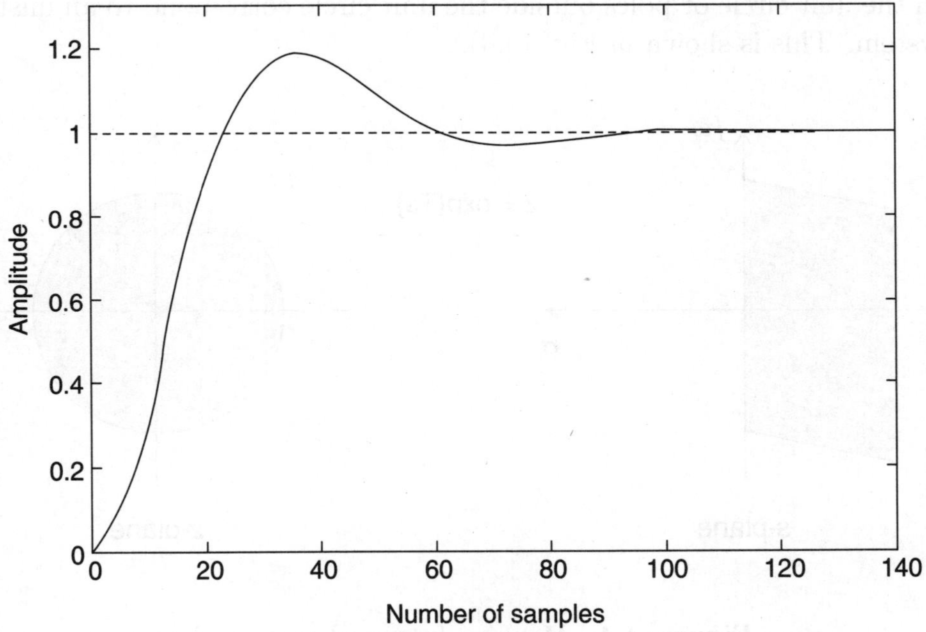

Figure 4.3: Step response of the system: Example 4.3.

4.3.1 Stability and Pole Locations

Recall the relationship between the Laplace operator, $s = \sigma + j\omega$, and the z-operator given by

$$z = e^{Ts} = e^{\sigma T} e^{j\omega T}.$$

And, the magnitude that is given by

$$|z| = e^{\sigma T}$$

suggests that

$$|z| = \begin{cases} < 1 & \text{if } \sigma < 0 \\ = 1 & \text{if } \sigma = 0 \\ > 1 & \text{if } \sigma > 0 \end{cases}$$

Hence, the points in the left half of the s-plane map into points inside the unit circle in the z-plane. The poles of a transfer function, $M(z)$, inside the unit circle, $|z| = 1$, correspond to a stable discrete time system. Poles on the unit circle correspond to a marginally stable system. Repeated poles on the unit circle or poles outside the unit circle correspond to an unstable system. This is shown in Fig. (4.4).

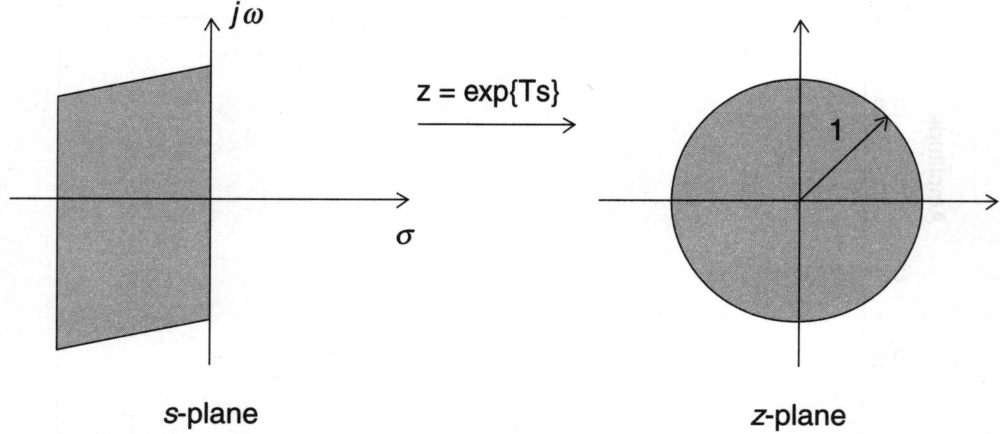

Figure 4.4: Mapping from s-plane to z-plane.

4.3.2 Modified Routh's Criterion

In this method, a transformation that is commonly called the w transformation is used to transform the inside of the unit circle in the z-plane into the left-half of another complex plane called the w-plane. The w transformation used for this purpose is also a bilinear transformation given by

$$z = \frac{1+w}{1-w}.$$

Let $w = \sigma + j\omega$. Now, the inside of the unit circles in the z-plane is

$$|z| = \left| \frac{1 + \sigma + j\omega}{1 - \sigma - j\omega} \right| = \frac{(1+\sigma)^2 + \omega^2}{(1-\sigma)^2 + \omega^2} < 1.$$

This yields

$$(1 + \sigma)^2 + \omega^2 < (1 - \sigma)^2 + \omega^2$$

or

$$\sigma < 0.$$

After this transformation is applied to the characteristics equation polynomial in z, the resulting equation can then be used to analyze stability using the Routh's criterion in the same manner as with the continuous time case.

Let the characteristic equation be

$$A(z) = a_0 z^n + a_1 z^{n-1} + a_2 z^{n-2} + \cdots + a_n = 0.$$

Using the transformation, we get

$$a_0 \left(\frac{1+w}{1-w}\right)^n + a_1 \left(\frac{1+w}{1-w}\right)^{n-1} + a_2 \left(\frac{1+w}{1-w}\right)^{n-2} + \cdots + a_n = 0.$$

By multiplying both sides by $(1 - w)^n$ and rearranging yield an equation of the following form

$$B(w) = b_0 w^n + b_1 w^{n-1} + b_2 w^{n-2} + \cdots + b_n = 0.$$

Now, it is possible to apply the Routh's criterion in the same manner as in continuous-time case.

Example 4.4: The characteristic equation of a closed-loop discrete time system is given by

$$45 z^3 - 117 z^2 + 119 z - 39 = 0.$$

Using the w transformation, we get

$$45 \left(\frac{1+w}{1-w}\right)^3 - 117 \left(\frac{1+w}{1-w}\right)^2 + 119 \left(\frac{1+w}{1-w}\right) - 39 = 0.$$

By multiplying both sides by $(1 - w)^3$ and rearranging yield

$$w^3 + 2w^2 + 2w + 40 = 0.$$

The Routh array for this system is

$$\begin{array}{c|cc}
w^3 & 1 & 2 \\
w^2 & 2 & 40 \\
w^1 & -18 & 0 \\
w^0 & 40 & 0
\end{array}$$

Since there are two sign changes in the first column, there are two roots of the characteristics equation in the right-half of the w-plane. This implies two roots outside the unit circle in the z-plane. Hence, the system is unstable.

Example 4.5: Consider the unity feedback sampled-data system given in Fig. (4.5)

where

$$G_p(s) = \frac{1}{s(s+4)} = \frac{1}{s^2 + 4s}$$

and $K > 0$ is an adjustable parameter.

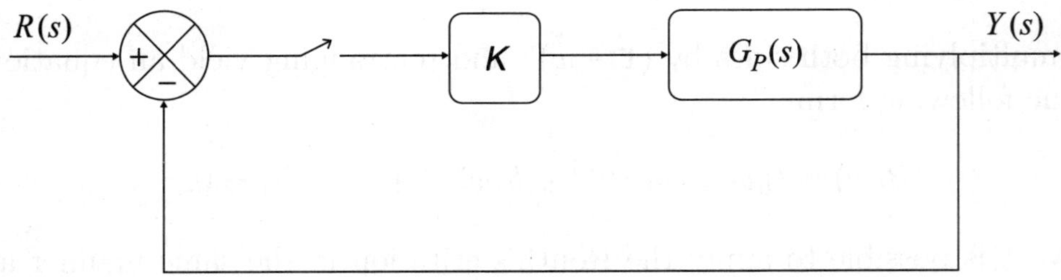

Figure 4.5: System for Example 4.5.

By applying Routh's criterion for the continuous system characteristic equation

$$1 + \frac{K}{s(s+4)} = 0$$

one may observe that the system without the sampler is stable for all values of K.

Let us study the effect of finite sampling on system closed-loop stability. Assume $T = 0.25$ s and that the impulse sampler is not followed by a hold circuit. Using the Matlab code

```
>> num = [1];
```

$$\begin{array}{c|cc} s^2 & 1 & K \\ s^1 & 4 & 0 \\ s^0 & K & \end{array}$$

```
>> den = [1 4 0];
>> T = 0.25;
>> [numd, dend] = c2dm(num, den, T,'imp');
>> printsys(numd, dend,'z')
```

we get

$$G_p(z) = \frac{0.158z}{z^2 - 1.3679z + 0.3679}.$$

Thus, the characteristic equation of the closed-loop system is

$$1 + KG_p(z) = z^2 + (0.158K - 1.368)z + 0.368 = 0.$$

By applying w-transformation and rearranging the terms yield

$$0.158Kw^2 + 1.264w + (2.736 - 0.158K) = 0.$$

The Routh array for this system is

$$\begin{array}{c|cc} w^2 & 0.158K & (2.736 - 0.158K) \\ w^1 & 1.264 & 0 \\ w^0 & (2.736 - 0.158\,K) & 0 \end{array}$$

For stability

$$K > 0$$

$$2.736 - 0.158K > 0 \ \text{ or } \ K < 17.316$$

yield $0 < K < 17.316$ to be the range for which the system is stable. By following the same lines as above for different values of T, one may observe that the range for which the system is stable improves when T is decreased. Generally, smaller the value of T, the closer the discrete system approximates the original system. Stability improves as the sampling rate, $\frac{1}{T}$, increases.

4.4 Design of Digital Control Systems

The control system should be designed to meet the design specifications. In digital implementation of the controllers, the mathematical equations of the control algorithm should be generated to be implemented on a digital processor such as a microcontroller or a DSP. There are two widely practiced approaches to the design of digital controllers.

1. Design as an analog controller and then transform to the digital domain by the use of a transformation such as bilinear transformation. This approach is generally valid for systems with high sampling frequencies and hence small sampling periods.

2. A direct approach that is valid for any sampling frequency is to design in the digital domain. Here, the continuous plant equations are transformed to the z-domain and the closed-loop system is analyzed in the z-domain.

Finally, the z-domain expressions are inverse z-transformed into difference equations that can be implemented on a digital processor.

The first approach is often preferred when the designer is familiar with analog design techniques. This approach is less acceptable when the sampling frequency is small. Further, it does not take into account the samplers and hold devices. Bilinear transformation with frequency pre-warping is one of the most commonly used analog-based design techniques. Frequency pre-warping is the method used to correct the distortion that takes place in mapping s-domain to the z-domain. This method matches the single most important critical frequency in the analog domain and the digital domain. This is done by replacing each s in the analog transfer function with

$$\frac{\omega_1}{\omega_2} s \tag{4.22}$$

where ω_1 is the frequency to be matched in the digital transfer function and

$$\omega_2 = \frac{2}{T} \tan(\omega_1 T/2). \tag{4.23}$$

Most of the classical analog design techniques only requires simple modifications before they can be used in the digital domain so that a direct digital controller design method can be adopted. With the direct approach, however, the approximations and limitations that arise from s to z-domain transformation are eliminated.

4.4.1 Root Locus Method

The root locus of a discrete system is a plot of the locus of the roots of the characteristic equation

$$1 + G(z)H(z) = 0$$

in the z-plane as a function of the loop gain, K. This method allows us to position the roots of the characteristic equation into desired locations in the z-plane by adjusting the loop gain, K, or by adding open-loop poles and/or zeros. The closed-loop system will remain stable provided the loci remains within the unit circle. Even though manual construction of the root locus by following similar steps to the analog case is possible, in this book, we will only resort to Matlab based realization and analysis of the root locus.

Example 4.6: Consider the unity feedback sampled-data system

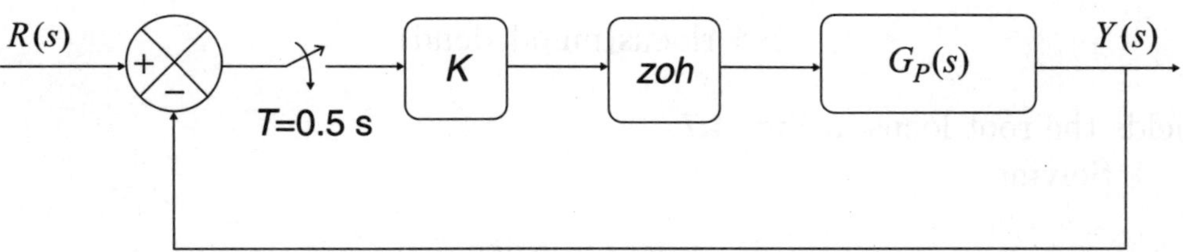

Figure 4.6: Sampled data system for Example 4.6.

given in Fig. 4.6 where

$$G_p(s) = \frac{1}{s(s+2)}$$

and $K > 0$ is an adjustable parameter.

In order to obtain the open-loop transfer function, consider

$$G(s) = KG_{zoh}(s)G_p(s) = K\frac{1 - e^{-Ts}}{s}\frac{1}{s(s+2)}.$$

Taking partial fraction expansion yields

$$G(s) = K(1 - e^{-Ts})\left\{\frac{-0.25}{s} + \frac{0.5}{s^2} + \frac{0.25}{s+2}\right\}.$$

Taking the z-transform, we get

$$G(z) = 0.25K(1 - z^{-1})\left\{-\frac{z}{z-1} + \frac{2Tz}{(z-1)^2} + \frac{z}{z-e^{-2T}}\right\}$$

whence substituting $T = 0.5$ s and simplifying yield the open-loop transfer function as

$$G(z) = K\frac{0.0952z + 0.066}{z^2 - 1.368z + 0.368}.$$

Alternatively, the following Matlab code can be used

```
>> num = [1];
>> den = [1 2 0];
>> T = 0.5;
>> [numd, dend] = c2dm(num, den, T,'zoh');
>> printsys(numd, dend,'z')
```

Now, the command

```
>> rlocus(numd, dend)
```

yields the root locus in Fig. 4.7.

Solving

$$z^2 - 1.368z + 0.368 = 0$$

yields the open-loop poles as

$$z = 1.0 \quad \text{and} \quad 0.368.$$

The open-loop zero is

$$0.0952z + 0.066 = 0$$
$$z = -0.717.$$

Two loci start from these open-loop poles. One of them ends at the open-loop zero $z = -0.717$ and the other at infinity. It can be seen from Fig. 4.7 that the root loci form a circle that is usually the case for second order systems. Obviously, the system is unstable if K is too large. The K values for marginal stability of the values of K at the boundary of the unit circle can be found using modified Routh's criterion as

$$K = 9.58.$$

Jury test another method that can be used as an alternative.

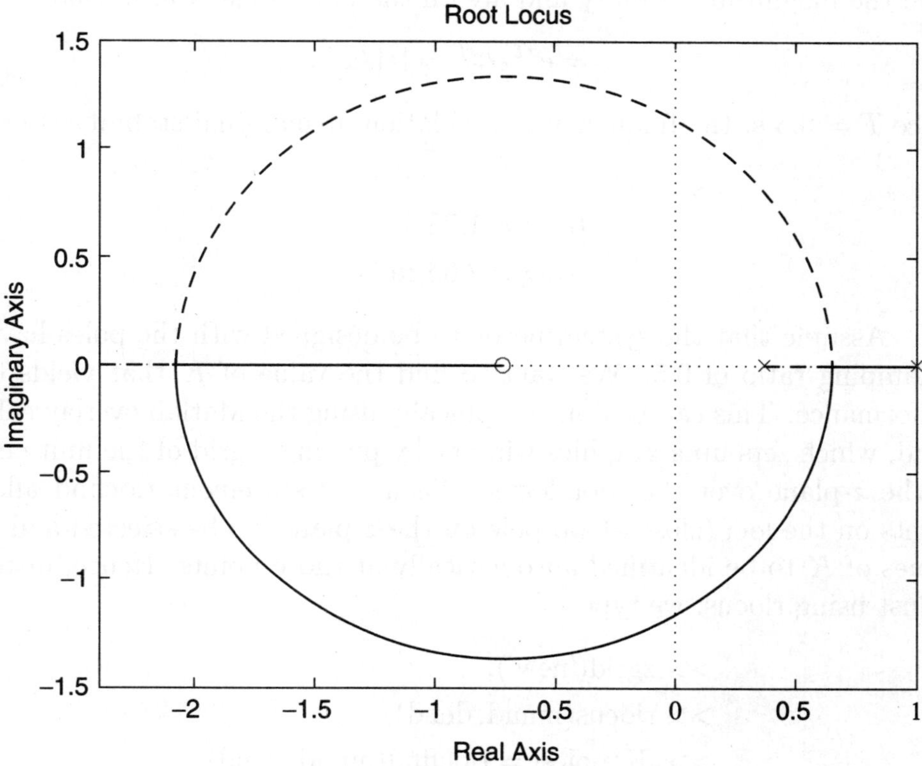

Figure 4.7: Matlab plot for Example 4.6.

The characteristic equation is

$$1 + G(z) = 0$$

$$1 + K\frac{0.0952z + 0.066}{z^2 - 1.368z + 0.368} = 0$$

$$z^2 + (0.092K - 1.368)z + (0.368 + 0.066K) = 0.$$

Substituting $K = 9.58$ into the characteristic equation gives

$$z^2 - 0.487z + 1 = 0$$

or

$$z = 0.244 \pm j0.97.$$

These roots that can also be expressed as

$$z = 1.0 \angle \pm 1.33 \text{ rad.}$$

have the magnitude of unity and are on the unit circle. Recall that

$$z = e^{\sigma T} e^{j\omega T} = |z|\angle \omega T.$$

Since $T = 0.5$ s, the frequency of oscillation at marginal stability is given by

$$0.5\omega = 1.33$$
$$\omega = 2.66 \text{ rad/s}.$$

Assume that the system needs to be designed with the poles having a damping ratio of 0.5. We want to find the value of K that yields this performance. This can be done graphically using the Matlab overlay called zgrid, which sets up a graphics window by putting a grid of the unit circle in the z-plane over the root locus. Then, the statement rlocfind allows points on the loci (closed-loop pole on the z-plane) to be selected and the values of K to be identified automatically at those points. Hence, instead of just using rlocus, we type

> >> zgrid('new');
> >> rlocus(numd, dend);
> >> [K, poles] = rlocfind(numd, dend)

and Matlab provides the plot shown in Fig. 4.8. The command, zgrid, creates an unit circle together with constant natural frequency and damping within the unit circle. In Fig. 4.8, we can position the cross-hairs at any point and get the closed-loop pole values and the values of K at that point. For instance, at the point where the root locus crosses the unit circle, we get $K = 9.668$ and poles at $0.2393\pm j0.9743$ that are very close to those calculated by the Routh's criterion. Locating $\zeta = 0.5$ line as the fifth line up from the real axis in the first quadrant, we position the cross-hairs where that line crosses the root locus. Here, Matlab indicates a value of $K = 2.1$ and the closed-loop poles at $0.5874\pm j0.4021$.

For a second-order system, it is generally desired that the closed-loop poles be in the first and fourth quadrants of the interior of the unit circle. For higher-order systems, we would often like a pair of dominant poles to be in the first and fourth quadrants of the interior of the unit circle.

4.4.2 Bode Plots

Bode plots which are the magnitude and phase angle plots against frequency of the open-loop transfer function of a system are used in frequency domain analysis methods. The Bode plot construction techniques

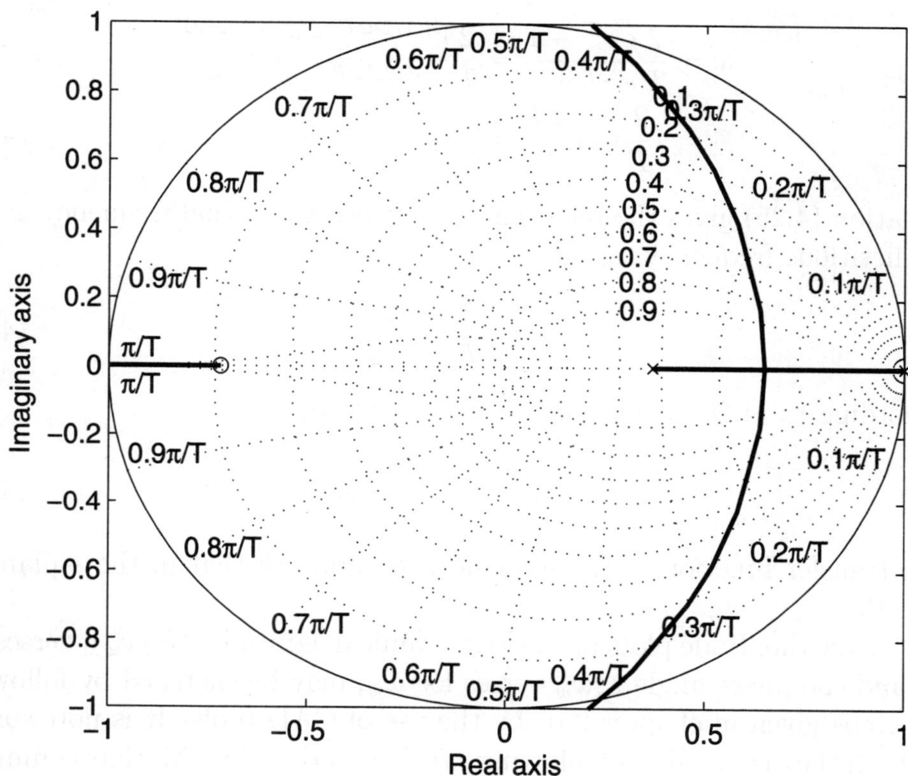

Figure 4.8: z-grid plot for Example 4.6.

and frequency domain analysis methods used for continuous systems can
be employed after transforming a transfer function $G(z)$ into $G(w)$ by the
use of the transformation

$$z = \frac{1 + (T/2)w}{1 - (T/2)w}.$$ (4.24)

Now, the transfer function in w may be treated as a conventional transfer
function and the familiar frequency domain techniques can be applied to
the design of discrete-time control systems. However, as shown shortly, this
approximate approach will be better valid for higher sampling frequencies.
If it can be afforded, 20 times the highest frequency in the system as the
sampling frequency is a very good choice.

In the frequency domain analysis, $s = j\omega$ and hence $z = e^{j\omega T}$ where

$1/T$ is the sampling frequency. Substituting $z = e^{j\omega T}$ in (4.24) and rearranging yield

$$w = \frac{2}{T}\frac{e^{j\omega T} - 1}{e^{j\omega T} + 1} = \frac{2}{T}\frac{e^{j(1/2)\omega T} - e^{-j(1/2)\omega T}}{e^{j(1/2)\omega T} + e^{j(1/2)\omega T}}$$

$$= j\frac{2}{T}\tan\frac{\omega T}{2} = j\omega_w. \tag{4.25}$$

Equation (4.25) gives the relationship between the actual frequency ω and the fictitious frequency ω_w as

$$\omega_w = \frac{2}{T}\tan\frac{\omega T}{2}. \tag{4.26}$$

According to (4.26), note that if ωT is small, then

$$\omega_w \approx \frac{2}{T}\frac{\omega T}{2} = \omega. \tag{4.27}$$

The transfer function in the w-plane will approach that in the s-plane as $T \to 0$.

Now the, Bode plots or the logarithmic magnitude $|G(j\omega_w)|$ verses log ω_w and the phase angle $\angle j\omega_w$ verses log ω_w, may be sketched by following the steps given in chapter 2 or by the use of CAD tools. It is noteworthy that Matlab is capable of plotting $G(z)$ directly. The Matlab command dbode(num, den, T) will calculate and plot Bode plots for any given G(z). Hence, $G(w)$ approach to design of discrete time systems is only attractive if the Bode plots are sketched manually.

Matlab uses a better approximation for $z = e^{j\omega T}$ than w-transformation given in (4.24). Difference between $G(z)$ and $G(w)$ Bode plots is smaller for smaller values of T.

4.4.3 Digital PID Control

In chapter 2, we discussed the PID controller, Ziegler-Nichols methods of gain tuning, and its analog implementation using operational amplifier circuits. Discrete time version was also briefly discussed. In computer implementation of PID control as well, Ziegler Nichols method can be used to design the controller. The control law is then converted to the difference equation and implemented by programming a DSP chip.

In Ziegler-Nichols method based PID controller design in the discrete-time domain, controller is normally designed as an analog controller and then transformed to the digital domain by the use of bilinear transformation. Bilinear transformation can be used to transform continuous domain

PID transfer function, (2.21), to the discrete domain as

$$G_C(z) = \frac{Az^2 + Bz + C}{z^2 - z} \tag{4.28}$$

where

$$A = k_P + k_I T + \frac{k_D}{T}, \ B = -k_P - 2\frac{k_D}{T}, \text{ and } C = \frac{k_D}{T}. \tag{4.29}$$

The gains k_P, k_I, k_D obtained through Ziegler-Nichols method and the value of T are used to calculate the constants A, B, and C in (4.28).

If the plant transfer function, $G_P(z)$, is available, one can resort to much convenient graphical methods such as root locus method or Bode plots to obtain k_P at marginal stability, $k_P = k_{ZN}$, and the frequency of oscillation, ω_{ZN}, by the use of Matlab. Since, $G_C(z)$ is also required, the following form of the PID transfer function may be used.

$$G_C(z) = k_P + k_D\frac{z-1}{Tz} + k_I\frac{Tz}{z-1}. \tag{4.30}$$

Now, the open-loop transfer function, $G_C(z)G_P(z)$, is available to carry out analysis using graphical methods. The optimal gains obtained using Ziegler-Nichols method can then be sued to calculate the constants A, B, and C in (4.28). The difference equation of the PID transfer function can now be obtained using (4.28) before implementing the control algorithm on a DSP chip.

4.4.4 State Variable Methods

For most lower-order systems, classical methods such as Bode plots or root-locus method are very much adequate and often preferred. However, as discussed in chapter 3, relatively modern design techniques such as state-space design are more sophisticated and can be used for complex higher-order systems. The discussion given in this section will be brief because of the strong analogy between discrete-time and continuous-time cases.

The state variable description of a linear discrete-time system is

$$\boldsymbol{x}((k+1)T) = F\boldsymbol{x}(kT) + G\boldsymbol{u}(kT) \tag{4.31}$$
$$\boldsymbol{y}(kT) = H\boldsymbol{x}(kT) \tag{4.32}$$

where $\boldsymbol{x}(kT) = [x_1(kT) \ x_2(kT) \ \cdots \ x_n(kT)]^T \in \Re^n$ is the n-dimensional vector of states and $\boldsymbol{u}(kT) \in \Re^q$ is the vector of inputs. $\boldsymbol{y}(kT) \in \Re^p$ is the vector of outputs. F, G, and H are matrices of suitable dimensions. Noise terms have been ignored. Clearly, (4.31) is a set of first order difference equations stacked together to form a matrix equation.

Typically, (4.31) and (4.32) equations are obtained by discretizing the state-space equations originally given in the continuous-time form. How to transform original state equations to the discrete domain is worth looking at. The solution of the continuous-time state equation (3.15) was obtained in chapter 3 as

$$\boldsymbol{x}(t) = e^{A(t-t_0)}\boldsymbol{x}(t_0) + \int_{t_0}^{t} e^{A(t-\tau)}Bu(\tau)d\tau. \tag{4.33}$$

Recall that we considered SISO systems only for the manual manipulation of these equations for convenience. Consider a time interval of T and let the initial time be $t_0 = kT$. Assuming a zeroth-order hold, we can write

$$u(t) = u(kT); \ kT \le t < (k+1)T \ \forall \ k.$$

Now, from (4.33), we get

$$\boldsymbol{x}(t) = e^{A(t-kT)}\boldsymbol{x}(kT) + \left\{\int_{kT}^{t} e^{A(t-\tau)}Bd\tau\right\}u(kT); \ kT \le t < (k+1)T. \tag{4.34}$$

At $t = (k+1)T$, the response settles at $\boldsymbol{x}((k+1)T)$ before the application of the input $u((k+1)T)$, and hence

$$\boldsymbol{x}((k+1)T) = e^{AT}\boldsymbol{x}(kT) + \left\{\int_{kT}^{(k+1)T} e^{A[(k+1)T-\tau]}Bd\tau\right\}u(kT) \tag{4.35}$$

$$= F\boldsymbol{x}(kT) + Gu(kT). \tag{4.36}$$

Let $\sigma = (\tau - kT)$. Now

$$G = \int_{0}^{T} e^{A(T-\sigma)}Bd\sigma.$$

Letting $\alpha = T - \sigma$, we get

$$G = \int_{0}^{T} e^{A\alpha}Bd\alpha.$$

If $\boldsymbol{x}(t)$ (or $y(t)$) is to be obtained between two consecutive sampling instants, one can obtain $x(kT)$ for any k from (4.34) and then use (4.36).

It is direct to obtain the equation of the sampled output $y(kT)$ by substituting $t = kT$ in the output equation (3.16) of the original continuous-time state space model. Hence, F, G, H matrices of the equivalent discrete-time system, (4.31), (4.32), can be obtained from the original continuous-time system, (3.78), (3.16), as

$$F = e^{AT} = \phi(T) \tag{4.37}$$

$$G = \int_0^T e^{A\alpha} B \, d\alpha = \int_0^T \phi(\alpha) B \, d\alpha \tag{4.38}$$

$$H = C \tag{4.39}$$

where $\phi(t)$ is the state transition matrix defined in (3.26).

Example 4.7: Consider the continuous-time system characterized by

$$\dot{x}(t) = \begin{bmatrix} 0 & 0 \\ 30 & -30 \end{bmatrix} x(t) + \begin{bmatrix} 1 & 1 \\ 1 & -1 \end{bmatrix} u(t)$$

and

$$y(t) = \begin{bmatrix} 1 & 1 \\ 1 & 1 \end{bmatrix} x(t).$$

To obtain the equivalent discrete-time state equations of the system, we first obtain the state transition matrix, $\phi(t)$.

$$[sI - A]^{-1} = \frac{1}{s(s+30)} \begin{bmatrix} s+30 & 0 \\ 30 & s \end{bmatrix} = \begin{bmatrix} 1/s & 0 \\ a/s(s+30) & 1/(s+30) \end{bmatrix}$$

$$= \frac{1}{30s} \begin{bmatrix} 30 & 0 \\ 30 & 0 \end{bmatrix} - \frac{1}{30(s+30)} \begin{bmatrix} 0 & 0 \\ 30 & -30 \end{bmatrix}$$

and

$$\phi(t) = \mathcal{L}^{-1}[(sI - A)^{-1}] = \begin{bmatrix} 1 & 0 \\ 1 - e^{-30t} & e^{-30t} \end{bmatrix}.$$

Hence

$$F = e^{AT} = \phi(T) = \begin{bmatrix} 1 & 0 \\ 1 - e^{-30T} & e^{-30T} \end{bmatrix}.$$

Further

$$G = \int_0^T \phi(\alpha) B d\alpha = \int_0^T \begin{bmatrix} 1 & 1 \\ 1 & 1 - 2e^{-30\alpha} \end{bmatrix} d\alpha$$

$$= \begin{bmatrix} T & T \\ T & T + \frac{1}{15}(e^{-30T} - 1) \end{bmatrix}.$$

If the sampling period, T, is known, the state-space representation of the equivalent discrete-time model can now be obtained. Note that as $H = C$, it is direct to obtain the output equation.

Example 4.8: Consider the following difference equation of a discrete-time system.

$$y(k + 2) + b_1 y(k + 1) + b_2 y(k) = a_1 u(k + 1) + a_2 u(k) \tag{4.40}$$

where b_1, b_2, a_1, a_2 are the coefficients which are constants if the system is time invariant.

Taking z-transform of the both sides and setting zero initial conditions, we get the transfer function of the system as

$$\frac{Y(z)}{U(z)} = \frac{a_1 z + a_2}{z^2 + b_1 z + b_2}. \tag{4.41}$$

In order to obtain the state variable form, the states may be defined as

$$x_1(k) = y(k) \tag{4.42}$$
$$x_2(k) = x_1(k + 1) - a_1 u(k) \tag{4.43}$$

whence substituted in (4.40), yields the state equations as

$$x_1(k + 1) = x_2(k) - a_1 u(k) \tag{4.44}$$
$$x_2(k + 1) = x_1(k + 2) - a_1 u(k + 1)$$
$$= -b_2 x_1(k) - b_1[x_2(k) + a_1 u(k)] + a_2 u(k)$$
$$= -b_2 x_1(k) - b_1 x_2(k) - (a_1 b_1 - a_2)u(k). \tag{4.45}$$

Equations (4.44) and (4.45) represent the state equations of the system with

$$F = \begin{bmatrix} 0 & 1 \\ -b_2 & -b_1 \end{bmatrix} \text{ and } G = \begin{bmatrix} -a_1 \\ -(a_1 b_1 - a_2) \end{bmatrix}.$$

4.4.4.1 Canonical forms   As in the continuous case, the canonical forms are important. A system representation may be transformed into one that is controllable/observable. For SISO systems, the CF and OF can be obtained by reading coefficients from the system transfer function. If the F, G, H matrices of an arbitrary state-space representation of a SISO system is given, the transfer function (see equation (4.46)) can be obtained by taking the Laplace transform of the state and output equations and assuming zero initial conditions as follows.

$$\frac{Y(z)}{U(z)} = H(zI - F)^{-1}G = \frac{b_{n-1}z^{n-1} + \cdots + b_1 z + b_0}{z^n + a_{n-1}z^{n-1} + \cdots + a_1 z + a_0}. \qquad (4.46)$$

Canonical form matrices, F_c, F_o, G_c, etc. can be obtained following the same procedure as that used in the continuous case. Also, recall the method discussed in chapter 3 to transform state-space representations into CF and OF without needing the transfer function. Same procedure is valid for the discrete-time case as well.

4.4.4.2 Solution of the state equations   Solution of the difference equations (4.31) for $\boldsymbol{x}(k)$ is analogous to the differential equation solution for $\boldsymbol{x}(t)$. It can be shown that

$$\boldsymbol{x}(k) = F^k \boldsymbol{x}(0) + \sum_{n=0}^{k-1} F^{k-n-1}Gu(n) \quad k \geq 0 \qquad (4.47)$$

where

$$F^k = z^{-1}[I - z^{-1}F]^{-1} \qquad (4.48)$$

is the discrete state-transition matrix. The output response can be obtained as $\boldsymbol{y}(k) = H\boldsymbol{x}(k); \; k \geq 0$.

Note that the solution of the state equation, in both continuous and discrete cases consists of two components, namely, the zero input response and the zero state response. The first term depends purely on the initial condition and the second on the input alone. This is a general property of linear systems. However, in transfer function analysis, for instance, only the zero state response is of interest.

4.4.5 Controllability and Observability

As in the continuous case, controllability test for discrete-time systems can help find out what conditions the system needs to satisfy to ensure its states can be steered to any desired final state, $\boldsymbol{x}(k_f)$, from an initial state, $\boldsymbol{x}(0)$, by some unconstrained control, $u(k)$, in a finite number of sampling instances.

For instance, if any of the state variables is independent of the control input, $u(k)$, such a state variable is simply uncontrollable. Quite similar to the continuous case, for a discrete-time system to be completely state controllable, it can be shown that the *controllability matrix*, $\mathcal{C}(F, G)$, defined as

$$\mathcal{C}(F, G) = [G \; FG \; \ldots F^{n-1}G] \tag{4.49}$$

must have rank, n. $F \in \Re^{n \times n}$ is the system matrix and n is the order of the system. The proof of this can be found in certain literature and is omitted in this book. As in the continuous case, controllability of a discrete-time system only depends on F and G, and is independent of the output matrix, H. This result holds for SISO and MIMO systems.

4.4.5.1 Observability   A discrete-time system is said to be observable or completely state observable, if for a given input, $u(k)$, the knowledge of the output $y(k)$ and input $u(k)$ sequences over a finite number of steps are sufficient to determine every state $x(0)$. The test for observability is quite similar to that of the continuous case. The system represented by F, G, H is completely state observable if and only if its observability matrix defined by

$$\mathcal{O}_c(G, H) = \begin{bmatrix} H \\ HF \\ \vdots \\ HF^{n-1} \end{bmatrix} \tag{4.50}$$

has rank n where n is the order of the system. Observability, like controllability, is a property of a system representation, not of a system per se. Transforming a system representation into OF, which is always observable is possible by following the similar lines discussed for the continuous case in chapter 3.

4.4.6 State-feedback Control

In pole-placement method of design in discrete-time domain, the desired control performance is achieved by placing the closed-loop poles at desirable locations inside the unit circle of the z-plane.

In the state feedback control of a linear plant, $x(k + 1) = Ax(k) + Bu(k)$, shown in Fig. 4.9, state variables, $x(k)$, are fed back to the input side through a feedback gain matrix, $K = [k_1 \; k_2 \cdots k_n]$. With familiar notations, $r(k)$ is the reference input.

The control, $u(t)$, is given by

$$u(k) = -Kx(k) + r(k) \tag{4.51}$$

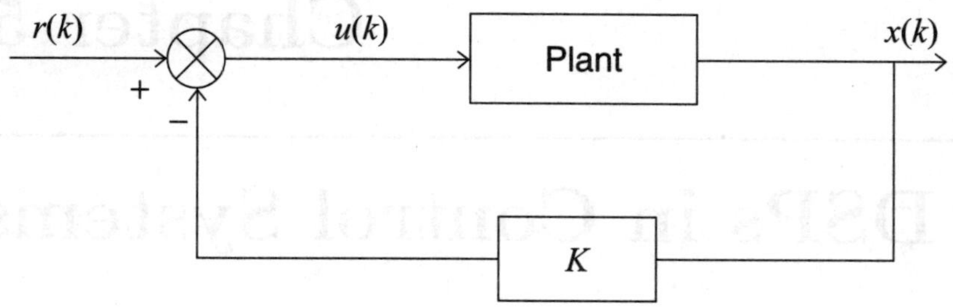

Figure 4.9: State feedback control.

whence substituted in the the plant equation yields

$$x(k+1) = Ax(k) + B[-Kx(k) + r(k)]. \tag{4.52}$$

This is rearranged to obtained the closed-loop system equation as

$$x(k+1) = (A - BK)x(k) + Br(k). \tag{4.53}$$

The design approach in this case is to determine the feedback gain, K, such that the desired performance characteristics are achieved by placing the closed-loop poles or eigenvalues of $A - BK$ given by

$$|zI - A + BK| = 0$$

at suitable locations on the z-plane.

Chapter 5

DSPs in Control Systems

5.1 Background

Owing to the rapid growth of microprocessor technology leading to high resolution, high processing power, and affordable costs, control algorithms nowadays are increasingly implemented through the use of embedded microcontrollers instead of analog devices. Microcontroller is a microprocessor specially designed for control applications. In these chips, speed, on-chip resources and software facilities are optimized for control applications. DSP (digital signal processor) is a specialized microprocessor designed specifically for heavy real-time computing. Some of them are special purpose microcontrollers specially designed for specific control applications. Examples are the DSPs available in the market for specific control applications that include automotive control (active suspension control, fuel injection control, anti-lock braking, etc.), high-speed motor control, control of industrial robot manipulators, and power converter control. However, most of the DSPs are suitable for a range of applications. For instance, Motorola DSP56K family has found home for many applications from motion control to audio.

Digital signal processing is the arithmetic processing of sampled signals to perform filtering, amplification, modulation, convolution, and transformations etc. Since digital computers use numerical methods to process digital signals, all these mathematical algorithms are accomplished through arithmetic operations. The processed digital signals are then converted back to continuous signals using digital-to-analog converters (DACs). To be able to understand the performance specifications of a microcontroller so that selecting and using one to implement control algorithms, it is important to understand the fundamentals of microcontrollers.

Common on-chip resources that commercial microcontroller systems consist of include memories, I/O ports, timers and conters, ADCs/DACs, PWM circuits etc. Memory devices or data storage commonly found are

RAM, ROM, PROM, and EPROM. A brief description of them is as follows.

1. Random access memory or data RAM is used in the working space during computations and data transfer. This returns any piece of data in a constant time regardless of its physical location and whether or not it is related to the previous piece of data.

2. Read only memory (ROM), programmable ROM (PROM), and errasable PROM (EPROM) are used for rapid source of information that are seldom or never be modified.

A microcontroller consists of one or many of signal converters from each type, ADCs and DACs. ADCs convert incoming analog signals into digital form at a given sampling period, T, and resolution. Typical resolution values are: 8 bits, 16 bits, and 32 bits etc. DACs convert outgoing signals that are essentially in digital form into analog form that the external world such as actuators understand.

The central processing unit (CPU) consists of arithmetic and logic unit (ALU) that performs arithmetic (add, subtract, negate, increment, and decrement) and logical (AND, OR, XOR, and NOT) operations, and a number of registers such as

1. Accumulators that undertake arithmetic operations.
2. Instruction register that holds the current instruction.
3. Data address register that holds the memory address of data.

A register is a small amount of storage available on the CPU whose contents can be accessed more quickly than storage available elsewhere.

DSPs usually have special architectures making them more suitable than general-purpose microprocessors for high-speed processing applications such as real time control. General-purpose microprocessors and low-end microcontrollers generally have a von Neumann architecture where instructions (program) and data are on the same bus. However, most DSPs use a Harvard architecture with data separated from the instructions to increase the speed. Rapid growth of VLSI technology has enabled parallel-processing architectures consisting of multiple processors. Multiple-processor DSPs are hence capable of dividing the signal processing task into several sub-tasks and executing them simultaneously yielding much higher speeds than a single-processor system.

Fig. 5.1 shows digital controller implementation based on ADSP 2102 chip from Analog Devices, Inc. Normally, a DSP performs control algorithms by executing the instructions from its program memory ROM. ROM may also contain other important information such as scaling factors. See, for instance, the implementation of PID controller using a fixed-point

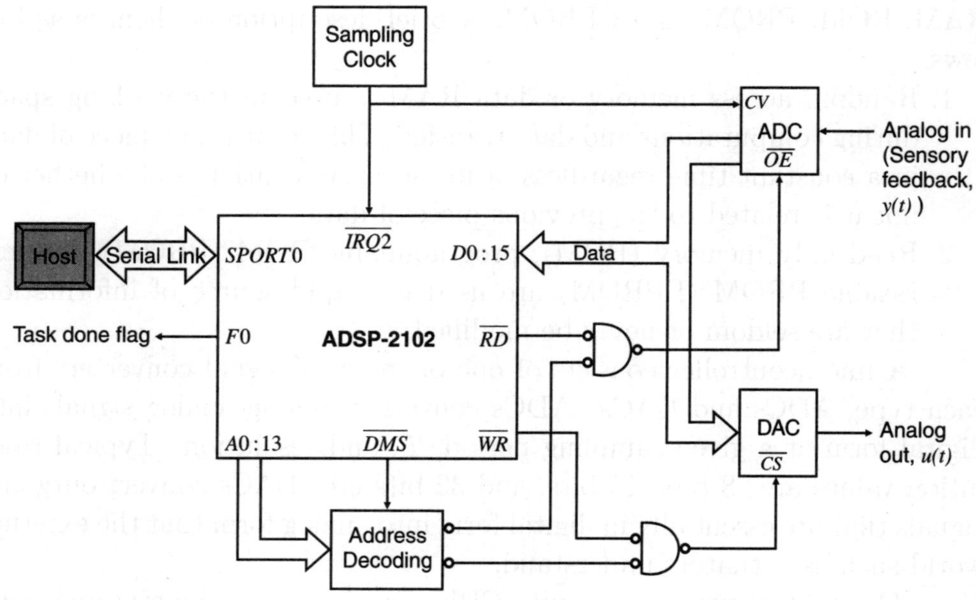

Figure 5.1: ADSP 2102 based digital controller. Courtesy of Analog Devices, Inc.

DSP in within this chapter. Processor automatically store incoming data and other intermediate variables in the program and data RAMs. A DSP chip normally accepts several external hardware interrupts. In the system given in Fig. 5.1, the external sampling clock drives one of the ADSP 2102 interrupts and, at regular intervals, the analog-to-digital conversions. The host computer uses other interrupts to notify the DSP of new commands, etc. A serial port is usually used to exchange commands and results with the host computer. The reference input, $r(t)$, is either received from the host or from an internal software routine running in parallel with the control algorithm. The DSP outputs the control command, $u(t)$, via an DAC. This signal is normally amplified before it is fed to the actuator. The DSP receives analog sensor feedback signals via ADCs. Data transfers between the processor and the converters are done over the data bus. In ADSP 2102, the converters are mapped into DSP's external data memory space using address decoding circuits allowing processor to access them as memory locations.

5.1.1 Selection of the Sampling Rate

In digital control, the continuous-time signals are sampled in time and quantized in magnitude. The processor acts on the digital signal. Hence, it is important to ensure that the signal that goes in represents the original continuous-time signal sufficiently accurately. Quantization error depends on the processor's word length that will be discussed shortly. Improper sampling can give rise to aliasing problem.

Let f_{max} is the highest frequency component of a continuous-time signal and f_s the sampling frequency. Shannon's sampling theorem states that the original signal is uniquely determined without introducing distortion only if

$$f_s \geq 2 f_{max}.$$

It is rare in practice to work near the limit $2 f_{max}$. A useful rule of thumb is to sample the signal with a sampling frequency of about ten times higher than f_{max}. When this result is applied to the field of digital control, the sampling period is usually selected above 10 to 20 times the system's bandwidth. If sampling occurs at a rate less than sufficient, we face an aliasing problem. To avoid aliasing effect, it is common to have an anti-aliasing filter typically with a cut-in frequency of about $0.5 f_s$ before the ADC. The higher f_s compared to f_{max}, the more closely the digital system resembles the analog system. Sampling frequency or sampling rate is a unique characteristic of a DSP and that determines the rate at which samples are produced.

5.1.2 Range and Round-off Error

The fundamental unit whereby information is represented in a digital computer is called a word. A word is a string of binary digits widely known as bits. Information is typically stored in one or more words of finite length. Primary logic units of a digital computer are on/off electronic components. Hence, the numbers on the computer are represented using a binary (1 or 0) base-2 (usually) system. For example, the binary number 101.1 in base-2 system is $(1 \times 2^2) + (0 \times 2^1) + (1 \times 2^0) + (1 \times 2^{-1}) = 5.5$ in the decimal or base-10 system. Due to the finite word length, digital computers can not represent some quantities exactly. For instance, irrational numbers such as π, e, or $\sqrt{7}$ can not be expressed by a finite number of significant figures. This may lead to erroneous results. Other implication of finite word length is the finite ranges of values that the computer can represent.

There are two major facets of errors occurring in digital computers.

1. Digital computers have size and precision limits in representing numbers.
2. Owing to the way computers perform arithmetic operations, certain numerical manipulations are highly sensitive to round-off errors.

Hence, we usually look for longer word length to minimize quantization errors and long register sizes to minimize truncation errors when selecting a microcontroller or a DSP for control applications. Quantization error occurs because some numbers can not be represented exactly. Minimum of 32-bit register size is often recommended to address the issue of truncation errors. For example, π can be realized using a 16-bit word size as 3.141593 while π would be expressed with 32-bit size with an increased accuracy as 3.14159265358979. In DSPs jargon, 32-bit precision is called *single precision*, 44-bit the *single extended precision*, and 64-bit the *double precision*.

Finite word length and subsequent dynamic range limitations can cause register overflow even leading to instability. There are finite ranges of values that the computer can represent. For example, with 16-bit word size, the range of integers is typically from -32,768 to 32,767 while with 32-bit size, this increases to -2,147,483,648 to 2,147,483,647. For fractional number or *floating-point numbers*, the corresponding ranges, respectively, are 10^{-38} to 10^{39} and 10^{-308} to 10^{308}. For instance, following Matlab code displays the smallest positive real number

$$>> \text{format long}$$
$$>> \text{realmin}$$
$$\text{ans} = 2.225073858507201e - 308.$$

Hence, Matlab can perform 32-bit precision. Numbers that are smaller than the above value are set to zero. If a value cannot be represented, next nearest value is produced. Difference between desired and actual value is the round-off error. For instance, subtraction of two very nearly equal numbers can give rise to large round-off error.

Truncation errors occur mainly due to the way in which computers perform arithmetic operations. In computing, truncation error occurs when the trailing digits are cut off or truncated after the maximum number of places allowed by the computer's level of accuracy. In practice, most numerical algorithms approximate desired solution with a finite number of arithmetic operations such as evaluating integral by quadrature, summing series using finite number of terms etc. Here, the difference between the true solution and the numerical approximation to the solution is called truncation error.

5.1.3 Fixed-point Versus Floating-point Processors

In fixed-point representation of data, real numbers can be represented by specifying some location in the word as the binary point. Binary notation can be extended to cover negative powers of 2. For example, binary number 110.101 is

$$1 \times 2^2 + 1 \times 2^1 + 0 \times 2^0 + 1 \times 2^{-1} + 0 \times 2^0 + 1 \times 2^{-3} = 6.625.$$

For most machines nowadays, real numbers are represented by floating-point format as

$$x = s \times M \times B^{(e-E)}$$

where s is the sign B is the base (usually 2, sometimes 16), M is the mantissa, e is the exponent, and E is the bias (usually 127). Manufacturers may use different number of bits for each of M and e, resulting in non-portable code. However, most manufacturers currently adopt IEEE standard where
    s is the first bit.
    Next 8 bits are e ($e = 255$ reserved for inf & NaN).
    Last 23 bits are M, expressed as a binary fraction, either 1.F, or, if $e = 0$, 0.F (in which case $E = 126$), where F is in base 2. For example

$$0\ 10000001\ 11100000000000000000000$$
$$= (+1)[2^{(129-127)}](1 + 0.5 + 0.125)$$
$$= 6.5.$$

An example is Motorola's DSP96002, which is a single-chip, 32-bit, IEEE floating-point DSP. Floating-points processors have a larger dynamic range compared with fixed-point processors and the finite word length effect is less significant.

Further, commercial floating-point processors are more expensive than fixed-point processors and have larger word length. Due to their dynamic range limitations in fixed-point processors, the signals are scaled with a scaling factor. Hence, these processors are comparatively difficult to work with. Floating-point DSPs are naturally preferred when rapid prototyping is needed as well as when the control problem is computationally intensive due to nonlinearities of the system, plant being higher-order MIMO, intense mathematics in control algorithms such as coordinate transformations etc. However, fixed-point processors have a range of control applications that include robot joint control, disk drives, and active suspension control of vehicles. Motorola's DSP56005 chip, for instance, is home for a range of control applications, especially in high-performance motor control and optical disk drives.

5.2 Single-chip DSPs

If the application does not require much of program ROM, data RAM, and many I/O ports, a single-chip microcontroller may serve the purpose. Microcontrollers such as MCS-51 family (Intel), HC 05 family (Motorola) include 8 and 16 bit processors with RAM, ROM, I/O ports, timer/counter, ADCs/DACs, and PWM circuits. However, with faster execution speed requirements and complexities in controller algorithms in high-performance control applications, a single processor microcontroller may not be able to meet the demands for speed, complexity, and flexibility.

DSPs generally adopt special hardware architectures to give high performance. The basic parts of a DSP architecture are the multiply-and-accumulate (MAC) unit, the program control unit (PCU), and the data/address bus interface units. Many important DSP algorithms such as FIR filters, fast Fourier transform (FFT), matrix operations (dot product, Gauss elimination, etc.) depend heavily on special arithmetic operations such as multiplication, multiplication with cumulative addition, and multiplication with cumulative subtraction performed by the MAC unit. The PCU of the DSP core coordinates the execution of program instructions and instructions for processing interrupts and exceptions.

Fig. 5.2 shows the DSP core of a 24-bit DSP56000 family (Motorola) chip. The PCU coordinates execution of instructions using three hardware blocks, the program address generator, the program decode controller, and the program interrupt controller to perform the following functions
- Fetch instructions
- Decode instructions
- Execute instructions
- Control hardware DO loops and REP
- Process interrupts and exceptions.

The PCU provides a system stack to store current register contents before executing the exception/interrupt handler program. These contents are hence restored when control returns to the current program. The program address generator contains the hardware needed for program address generation, system stack, and loop control. Program decode controller decodes the 24-bit instruction loaded into the instruction latch, generates signals for pipeline control, performs required data transfers between the data ALU and memory. Program interrupt controller arbitrates among interrupts requests and generates the appropriate interrupt vector address.

The device has a Harvard architecture allowing for data and instructions to be accessed within the same cycle. Four independent 24-bit data

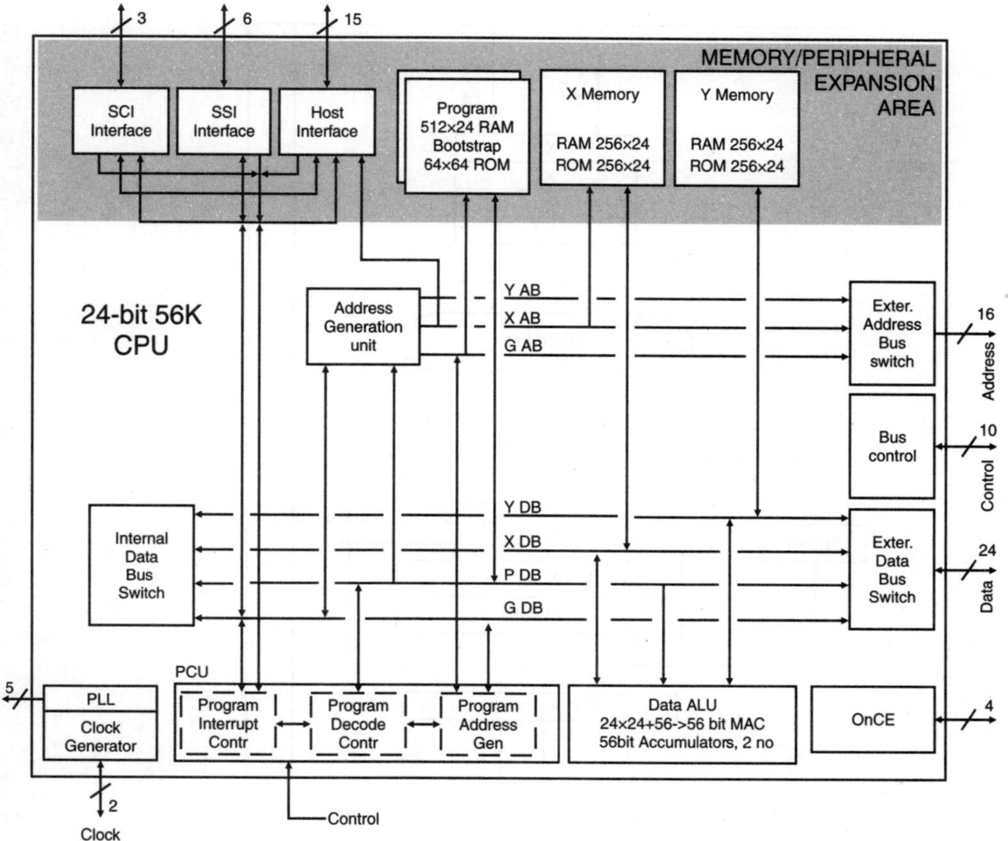

Figure 5.2: DSP56000 family DSP core diagram. Courtesy of Motorola, Inc.

buses are X-data buss (XDB), YDB, PDB, and global data bus (GDB). Three independent 16-bit address buses are XAB, YAB, and GAB. Address generation unit has 8 16-bit address registers, $R_0, R_1, \cdots, R_7$ that point to memory locations. 8 16-bit offset registers are added to the respective $R_n; n = 0, 1, \cdots, 7$ to form the final address and 8 16-bit modifier registers specify the type of arithmetic to update address register.

See Fig. 5.3. MAC and logic unit provides addition, subtraction, multiplication, logic functions, rounding, left and right shift, up and down scaling, limiting, bit testing and clearing. It gets up to 3 input operands from

- The memory
- The accumulators through feedback
- The multiplier.

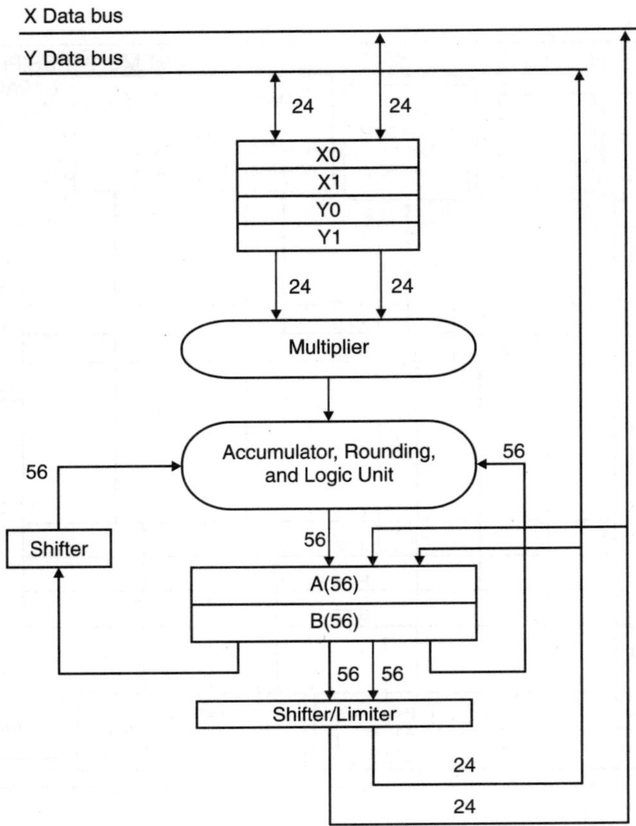

Figure 5.3: Data registers and MAC/ALU of DSP56000 core. X0, X1, Y0, Y1 are data input registers and A, B are accumulator registers.

Intermediate results are held in the accumulator registers, accumulator(A) or accumulator(B), of 56-bit word length. In general, longer the accumulator the less the quantization error will be. Left-shift and right-shift moves are important to perform multiplication and division by 2's, respectively.

Performance is usually measured in the number of cycles in which a DSP can accomplish basic calculations. Special architectural features of DSPs yield their corresponding performance specifications. Single-cycle instruction enables executing computationally hard control algorithms realtime, pipelined architecture controls high-bandwidth systems, Harvard architecture enables accessing data and instructions simultaneously.

Data and instructions are stored in RAMs. Fast access time for these RAMs is hence critical for many high-speed applications. In addition to on-chip RAMs, many DSPs give external memory expansion facility.

The DSP should be selected to suit the type of application. Some most important considerations are

- General arithmetic architecture
- Speed of performing arithmetic and other operations within a single cycle
- Amount of address space
- Single cycle instruction for high sampling rates
- Multi-stage pipelining to avoid pipeline problems for loops or for jumps
- Bus architecture for simultaneous access of data and instructions
- Long registers to minimize truncation errors
- Longer word length to minimize quantization errors
- Special on-chip resources such as sufficient number of on-chip I/Os, ADCs/DACs, PWM circuits. Depending on the task at hand, whether dedicated interface to certain buses, ability to communicate with other DSPs, etc. are required.

Often, the mathematical operations and other on-chip software resources required for control system applications are benchmarked by DSP chip manufacturers. They include

- Matrix manipulation
- Real and complex number multiplication
- Boolean algebra
- Pulse Width Modulators (PWM)
- Fast Fourier Transform (FFT)
- Finite impulse response (FIR), IIR, butterfly filters
- PID loop calculations
- Sine and arc-tangent tables
- n^{th}-order power series.

Motorola 56000 family DSPs use DSP56000 core. DSP56005 is a high-performance solution for many DSP and control applications, especially in high-performance motor control and optical disk drives. Its internal architecture yields the following features:

- Four 24-bit internal data buses and three 16-bit internal address buses for simultaneous accesses to one program and two data memories (Harvard architecture)
- Up to 25 million instructions per second (MIPS) 40 ns instruction cycle at 50 MHz
- Up to 150 million operations per second (MOPS) at 50 MHz

- Parallel 24×24-bit multiply-accumulate in 1 instruction cycle (2 clock cycles)
- Double precision 48×48-bit multiply with 96-bit result in 6 instruction cycles
- 56-bit Addition/subtraction in 1 instruction cycle
- 4608×24-bit on-chip program RAM and 96×24-bit bootstrap ROM, two 256×24-bit on-chip data RAMs, two 256×24-bit on-chip data ROMs containing sine and arc-tangent tables
- External memory expansion, bootstrap loading (from external data bus, host interface, or serial communications interface).
- A serial communication interface (SCI), synchronous serial interface (SSI), parallel host interface (HI), a 24-bit timer/event counter, and on-chip emulation (OnCE) port.

On-chip emulation (OnCE) port supports unobtrusive, processor speed-independent debugging. Host interface (HI) has direct memory access (DMA) support. Synchronous serial interface (SSI) is to communicate with codecs and synchronous serial devices. Serial communication interface (SCI) can be used for full-duplex asynchronous communications. Clock signals sequence the operation in DSPs. A serial port transmits and receives data one bit at a time. A synchronous serial port transmits a single-bit clock signal along with the data bit. A host port which is usually a parallel port communicates with other DSPs and microprocessors. A parallel port transmits and receives more than one bit at a time and hence provides a faster data transfer than a serial port at the cost of some additional pins.

Table 5.1 and Table 5.2 give details of some popular single-chip DSPs. Package types CQFP, PQFP, TQFP, and PGA stand for ceramic quad flat pack, plastic QFP, thin QFT, and pin grid array, respectively.

5.2.1 Programming, Device Evaluation and Debugging

Even though there are high-level utilities to program using high-level languages such as C, the most efficient way to program DSPs is with the assembly language. Hence, assembler/linkers and software simulators are usually provided by the DSP vendors to program and debug DSP algorithms. The assembler translates source files written in assembly language into machine language object files. The linker combines object files into an executable file.

Let us take a look at some of the development tools available from popular DSP vendors as examples. DSP56000 family chips, for instance, allow straightforward program development in either assembly language or a high-level language such as ANSI C. The software package DSP56KCCx is a

Table 5.1: Some popular single-chip DSPs

Model No	Word/Clock	Miscellaneous features
DSP56001 (Motorola)	24-bit, 33 MHz Fixed-point processor	24-bit data/instruction and 16-bit address 24×24-bit multiplication, 56-bit addition, 56-bit accumulator 24-bit(16-bit) external data(address) bus 1-cycle MAC, 132-pin CQFP/PQFP
DSP56005 (Motorola)	24-bit, 50 MHz Fixed-point processor	24-bit data/instruction and 16-bit address 24×24-bit multiply-accumulate 48×48-bit multiplication, two 56-bit accumulators, 24-bit(16-bit) external data (address) bus, 1-cycle MAC, 144-pin TQFP
DSP96002 (Motorola)	32-bit, 60 MHz Floating-point processor	32-bit data, instruction, and address Single-cycle 32×32-bit parallel multiplier IEEE standard 32-bit and 44-bit arithmetic 96-bit accumulator, 32-bit external data and address bus, 1-cycle MAC execution 2 preprogrammable ROMs, 223-pin plastic PGA or 240-pin CQFP package
ADSP2181 (Analog Devices Inc.)	16-bit, 40 MHz Fixed-point processor	24-bit instruction and 16-bit data 16K×16-bit data RAM 16K×24-bit program RAM 24-bit(14-bit) external data(address) bus 128-lead TQFP/PQFP

full ANSI C cross compiler that supports development of DSP56000 family applications in a stand-alone or mixed-language programming environment (C and 56000 assembly). The software consists of the C compiler, ANSI C libraries, assembler, linker, librarian, and GNU source level debugger. Like DSP56KCCx package, C compiler of the ADSP-21xx family development software is also based on the Free Software Foundation's GNU C compiler. The EZ-KIT Lite is a hardware/software development kit for the ADSP-21xx family. The assembler has an algebraic syntax that is easy to program and debug. The source code debugger allows programs to be corrected in the C environment. EZ-KIT Lite is an ADSP-21xx evaluation board with PC monitor software as well as assembler, linker, simulator, and runtime

Table 5.2: Some popular single-chip DSPs, contd.

Model No	Word/Clock	Miscellaneous features
ADSP21160 (Analog Devices Inc.)	32-bit, 100 MHz Floating-point processor	32/40-bit data, 48-bit instruction 600 MFLOPS, 2×32-bit IEEE float.-point computation units, 32 circ. buffers two 80-bit accumulators, 64-bit(32-bit) external data(address bus), 400-ball 27×27 mm PBGA package
TMS320C6201 (Texas Instruments)	16-bit, 200 MHz Fixed-point processor	32-bit instruction and 16-bit data 32-bit addr. space, 2×32k-byte blocks of data RAM, 64k-byte block of prog. memory, 40-bit accumulator, 8 circular-buffers, 27×27 mm 352-pin BGA pack.
TMS320C6701 (Texas Instruments)	32-bit, 120/ 150/167 MHz Floating-point processor	32-bit data and instruction, 2×32k-byte data RAMs, 64K-byte block of prog. memory, 1 GFLOPS, 8 circular buffers 40-bit accumulator 32-bit external memory interface 27×27 mm 352-pin BGA pack.

library of ANSI-standard and DSP-specific functions. The TMS320C6000 C/C++ compiler from Texas Instruments (TI) accepts C++ as well as ANSI C source code and produces assembly language source code for the TMS320C6000 devices. In addition to the assembler, TI development tool consists of an assembly optimizer that optimizes the code allowing achieve efficient code in a much shorter time. Such features are available with the development tools from many other vendors as well.

Software simulators that are provided by the DSP vendors are used for simulation of instruction sets of the device. Software simulators allow debugging without the target hardware, modification of register and memory contents. For instance, ADSP 21xx simulator provides an interactive instruction-level simulation with a reconfigurable user interface to display different portions of the hardware environment.

As the complexity of applications escalates, we are seeing an increased need for automated tools to aid in the development of DSP software. Code generation software automatically generates assembly code for a particular

DSP from the description control algorithm. An example is code generation tools available for the TMS320 DSPs. The techniques used to achieve
this may include high-level block-diagram-based modeling of the DSP application followed by the translation of block-diagram specifications into
efficient C programs using optimization techniques. The C programs are
then compiled into the machine code of the DSP using architecture-specific
methods. Use of code generation software may be useful when rapid prototyping is needed. However, as these tools have been designed to achieve
the best optimization possible for the entire code and not selected pieces,
inefficiencies may arise in a certain kernel of code.

Many DSP vendors provide the user with development hardware tools
such as hardware emulators that allow system-level integration, device evaluation, and hardware debugging. They allow user to perform single-step
or full-speed emulation with hardware break-points. An example is ADSP-
218x EZ-ICE emulator. It aids in the hardware debugging of ADSP-218x
systems. The emulator consists of hardware and host computer resident
software. The ADSP-218x integrates on-chip emulation support with a
14-pin interface and performs a range of functions including in-target operation, 20 breakpoints, single-step or full-speed operation, registers and
memory value be examined and altered, and C source-level debugging.

In the light of the ongoing discussion, the development tools for DSPs
can be summarized as in Fig. 5.4

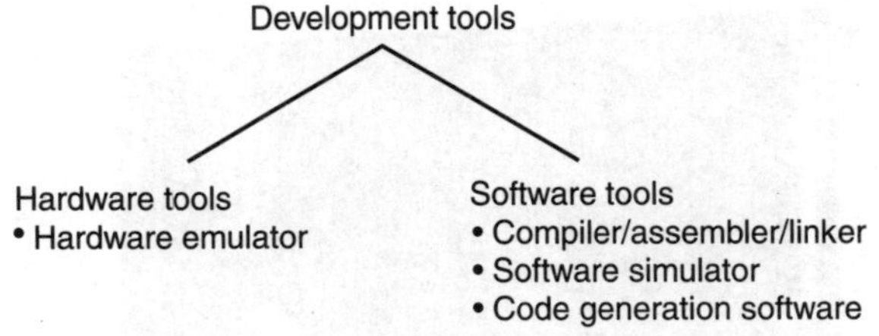

Figure 5.4: Development tools for DSP.

5.2.2 DSP-based Chip Sets and Control Boards

A chip set in contrast to a single-chip DSP is a set of multiple ICs that
perform specific key functions of a DSP. The key advantage is that with
a chip set, the designer can combine the individual ICs in a way that
best suits the specific application. An example is the motion control chip
set: AD2S90 resolver-to-digital converter, ADMC2001 motion processors,

and ADSP-2105 digital signal processor from Analog Devices, Inc. With this chip set, one can implement the sensory feedback, power converter control, and outer motion loop control for a range of motion control systems including AC servo motor control.

Controller boards are single-board systems with realtime hardware and I/O. These can be used to develop controllers for rapid prototyping. Such a board consists of a powerful floating-point processor, some slave DSP chips, as well as onboard peripherals and software resources. An example is DS1104 R&D Controller Board from dSPACE. This can be connected to a PC via PCI interface. Function models can be run and tested after configuring them graphically and inserting into Simulink block diagram.

Multi-axis motion control boards are used in embedded systems for industrial control, automation, and robotic applications. These motion control cards usually support multiple motor types, including DC brush, brushless DC, micro-stepping, and step motors. Examples are Prodigy-PCI and Magellan-PCI cards from Performance Motion Devices, Inc. that are available in 1, 2, 3, and 4-axis versions. The 16-axis motion control board, ADLINK PCI-8392 shown in Fig. 5.5, is used in high-end PC-based production machines. Such multi-axis applications demand absolutely synchronized motion for high speed and high accuracy systems. The board is based on the TI's TMS320C6711 DSP, which processes the synchronization.

Figure 5.5: DSP-based 16-axis motion control board. Courtesy of ADLINK Technology Inc.

5.3 Digital Implementation of Control Systems

5.3.1 PID Controllers

Since PID controller and its variants are heavily used in practice, DSP vendors often provide the guidelines on how to implement PID algorithms on their DSPs. They may sometimes include the assembly language subroutine that implements the PID algorithm. Implementing PID controllers on floating-point DSPs is straightforward. However, for fixed-point DSPs, some important guidelines need to be followed.

- In order to conform to the 16-bit (usually) fixed-point fractional number format, the gains, input and output variables must all be scaled down by the same factor. To this end, *a priori* knowledge of the range of the signals is needed. This also makes sure register overflows will not occur in the final stage of the multiply-accumulate operations. Results are eventually scaled up before being output to the controlled system.

- PID gain tuning is usually done with commercial CAD packages in higher precision arithmetic than 16 bits. For example, Matlab has 32-bit precision. System performance will be different when the gains are quantized to 16 bits and further scaled down. In applications requiring stringent PID specifications, careful simulations of quantization and scaling effects must be performed using the available programs and simulators from the DSP manufacturer.

5.3.2 n^{th} Order Digital Controllers

Digital filters can be classified according to the duration of the impulse response. The block diagram of a biquad 2nd order IIR (Infinite Impulse Response) filter section is given in Fig. 5.6 where a_1, a_2, b_0, b_1, b_2 are coefficients that determine the desired impulse response. The transfer function in the z-domain is given by

$$G(z) = \frac{U(z)}{E(z)} = \frac{b_0 + b_1 z^{-1} + b_2 z^{-2}}{1 + a_1 z^{-1} + a_2 z^{-2}}. \tag{5.1}$$

Hence, the corresponding difference equation of the biquad 2nd order IIR filter is

$$u(k) = b_0 e(k) + b_1 e(k-1) + b_2 e(k-2) - a_1 u(k-1) - a_2 u(k-2) \tag{5.2}$$

Standard 2nd order linear controller implementation is directly analogous to IIR filter implementations. In DSPs-based implementation, higher

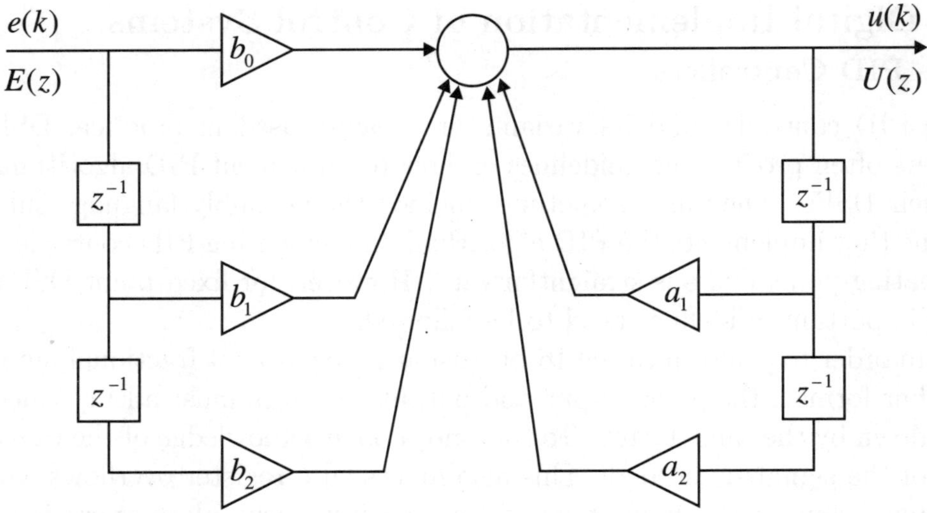

Figure 5.6: 2nd order biquad filter.

order controllers can be obtained either by cascading several biquad sections with appropriate coefficients or by direct form of implementation in which the controller is implemented as one complicated single section. IIR filters are also called recursive filters because the previous values of the outputs are used to obtain the current output. Hence, errors in previous outputs may accumulate. For fixed-point DSPs, the important guidelines mentioned for PID implementation may apply here as well. Biquad controller software routines often come as an important on-chip software resource with the DSPs.

5.3.3 Motor Control

Motor control systems are found in a wide range of applications from high-end industrial robotics to ordinary home appliances. High-performance servomotor control is needed in robotics whereas only adjustable-speed motor control may be required in an air conditioning system or a domestic washing machine. Adjustable-speed motor control is a feed-forward control system requiring no motor position or speed feedback. A sound background about the following classes of electric motors may be necessary to understand the important subtopic of servomotor control systems.

- *DC motors*: In a DC motor, the electromagnetic torque is produced by the interaction between the field flux established by the stator and the currents in the armature conductors. Field flux can be varied by varying the current in the field winding of field-wound DC motors. Hence, motor control can be achieved either by varying the torque producing current (armature control) or the field producing current (field control) or both. In permanent magnet (PM) DC motors, the field is produced by means of permanent magnets. PM DC motors are used in low power applications.

- *AC motors*: Induction motors and synchronous motors are the popular broad classes of AC motors. Squirrel cage induction motors are the workhorse of the industry and are more reliable, rugged, less expensive, and need less maintenance than DC motors. Induction motors are used both in adjustable-speed (fans, compressors) and servo applications (machine tools, robotics). Synchronous motors are also home for both servo (computer peripherals, robotics) and adjustable-speed (large fans, compressors) applications. PM synchronous motors of which the rotor is a PM are used in low power applications up to a few kilowatts.

- *Special-purpose motors*: Special types of motors include stepper motors, switch-reluctance motors, and piezoelectric motors.

- *Servomotors*: A servomotor is any of the aforementioned type of motor (DC, induction, synchronous, stepper, and so forth) that is designed to be used in servo applications where servomechanism is used. Servomechanism refers to as using automatic feedback control to correct the performance of a mechanism. Hence, a servomotor necessarily accompanies shaft position and/or speed measuring devices such as shaft-mounted resolvers and tacho-meters. A commercial servomotor often comes as a single package that comprises the motor, shaft position and/or speed sensors, power electronic motor drive and circuits, and reduction gears.

A servomotor control system typically has two cascaded control loops, namely, the outer motion control loop and the inner current control loop. The outer motion control loop works with the motion controller to control motor position and velocity, and requires motor shaft position or speed feedback or both, depending on the motion control algorithm. Sometimes, one of these motion parameters is numerically approximated or estimated using an observer based on the available measurement. The output of the motion controller is a demand for an increase or decrease in motor torque, which is fed to the inner current loop. Motor current-torque relationship (motor model) is used to convert the torque demands to current demands. The inner current loop generates the control signals for the PWM scheme, which then produces the switching signals for the power converter switching devices such as IGBTs or power MOSFETs. The on and off conduction periods of the switching devices are varied rapidly in a way the motor is supplied with the suitable currents to produce the desired output torque.

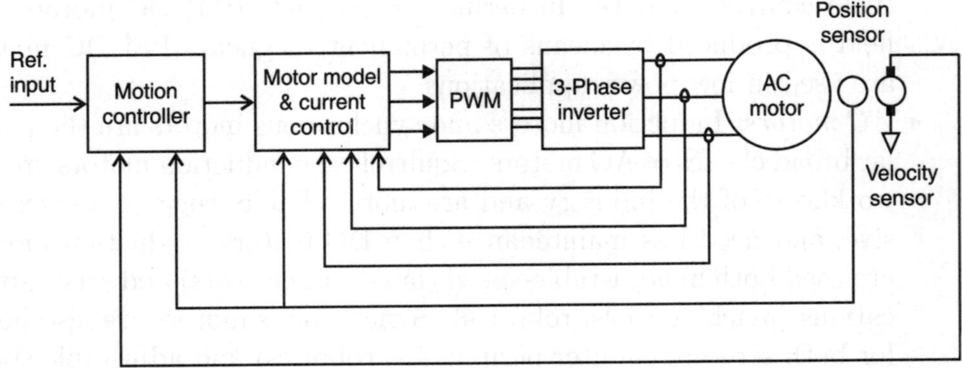

Figure 5.7: AC servomotor control system.

Fig. 5.7 shows an AC servomotor control system. AC servo control is more challenging in that the 3-phase currents are coupled and not independently controllable. Furthermore, torque-current and current-flux relationships are nonlinear. One way to simplify the control of the motor torque is to transform the measured stator currents to a reference frame synchronized to the rotor field.

Example 5.1: Vector control and field-oriented control are two commonly used methods for AC servo motor control. In these types of control, the stator current is split into DC equivalent torque producing component (I_q) and the field control component (I_d). See the reverse vector transform block of Fig. 5.9, which shows field oriented control of a PM synchronous motor using motion control chip set: AD2S90 resolver-to-digital converter, ADMC2001 motion processor, and ADSP-2105 fixed-point DSP from Analog Devices, Inc.. This stator current decomposition technique greatly simplifies the analysis, making it equivalent to analyzing separately-excited DC motors because in this case there are two independently controllable currents: the field current and the armature current. The current-loop controllers (eg. PI) produce two quadrature voltages, V_q and V_d, that can be used to force I_q to directly follow the torque demand and I_d to maintain the rotor field. A transformation is used to produce the control inputs to the PWM block. The outer motion control system produces the required motor torque current, I_q^*. The field reduction component, I_d^*, will be zero if field-weakening function is not effected. However, field weakening control can be effected by setting I_d^* values as a function of the motor speed, ω, and a wider motor speed range can be obtained.

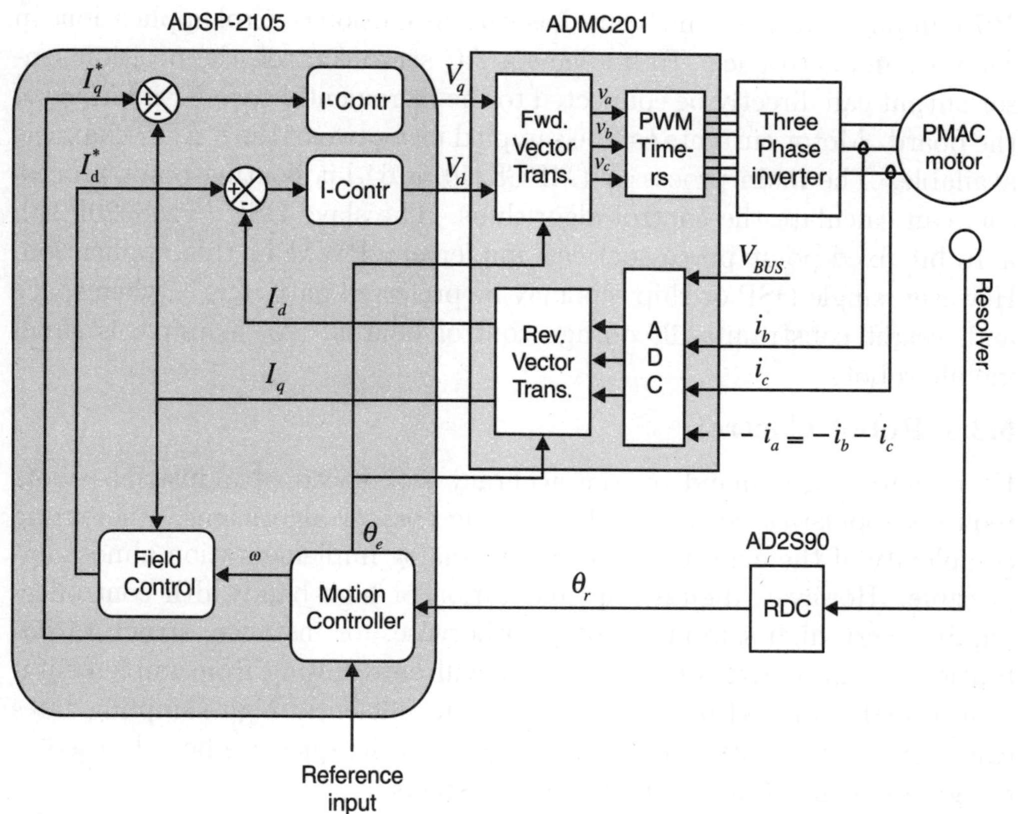

Figure 5.8: DSP-based AC servomotor control system.

ADMC201 samples the motor currents and passes them on as inputs to the vector transformation. The reverse transformation takes the current signals and the rotor electrical angle, θ_e, and calculates I_q, I_d quadrature current components. The two current controllers in the two current loops implemented in the DSP, calculate the quadrature voltage references V_q, V_d. The forward transformation block maps the the quadrature voltages into AC voltages within the stator reference frame. The DSP scales and then writes these results to the PWM block of the ADM201. The PWM timer block converts the digital inputs to pulse width modulated timing signals for the 3-phase inverter. The control loops are implemented in the ADSP-2105 DSP chip. ADMC201 chip, which interfaces the DSP to the 3-phase inverter can be connected directly to the DSP data and address buses. The AD2S90 resolver-to-digital converter can be connected to the DSP serial port.

An alternative may be to use controller boards. DS1104 R&D Controller Board from dSPACE, for instance, is a single-board system of 185× 107 mm physical size. The board has various motion control applications in robotics and aerospace. To implement AC servo control, the position sensor output can directly be connected to the incremental encoder interface of the board. Motor currents can be sampled using two of the 8 ADC channels available. The main processor (MPC8240, a 64-bit floating-point processor) can calculate the control algorithms. The slave DSP (TMS320F240, a 16-bit fixed-point processor) can implement PWM in this application. However, single DSP or chip sets may be preferred particularly when space and weight constraints discourage control boards. An example is small mobile robots.

5.3.4 Robot Control

Ever increasing demand on the accuracy and speed of industrial robots requires sophisticated and multi-variable control algorithms. Increasing complexity of the algorithms has made analog implementation almost impossible. However, digital implementation of high-bandwidth controllers requires very high sampling rates. Otherwise, for instance, structural vibrations of industrial robot arms that will be resulting from implementation of high-bandwidth controllers with insufficiently high sampling rates can degrade the control performance. DSP techniques are heavily used in design and control of variety of robot systems.

DSP systems provide an ideal solution to implement numerically intensive multi-variable control. If rapid prototyping is needed, control boards may be much preferred over implementing the system by optimally integrating several of commercial chips. However, if the space and payload restrictions are matters of concern, to develop a multi-DSP system using commercial off-the-shelf chips may be a better solution. Further, single chip DSPs consume very little power and smaller in size, which are prime concerns in small mobile robots. In such a case, the several subsystems of the robot may be implemented in several chips, which are then made to corporate in a multi-DSP system.

Example 5.2: The 32-bit TMS320F2812 single-chip DSP from Texas Instruments is used to implement the small robot system described in [11]. The DSP chip can implement processing of signals from multiple of sensors and control of multiple of motors of the small mobile robot. Two Mabushi 540 DC motors are used to power the drive wheels of a 4-wheeled robot system. Two 5.0A Motorola H-bridges with load current feedback are used to drive these motors. To make the robot motion estimation based on odometry, there are two optical encoders mounted on the wheel shafts. To realize obstacle avoidance, ultrasonic-based time-of-flight method measures the distance to the obstacles while the distance IR sensors measure the incident angle of reflected light. Bump sensors are tactile or touch sensors normally used to detect low obstacles that are not perceivable by other sensors. Two off-the-shelf servomotors (eg: high-torque hobby servo motors) may be used as the steering actuators to vary the steering angle of the front wheels in order to safely navigate the robot.

Four of the eight PWM blocks TMS320F2812 may be used to control the DC motors. The DC motors are driven by comparing the unit outputs PWM1,2, PWM3,4 via the H-bridge driver. One of the 4 Timers is used as the time base to generate PWM signal (eg: PWM frequency of 20 kHz). F2812 DSP has two built-in quadrature encoder pulse (QEP) circuits. The encoder readings of the driving wheel are easily obtained using the QEP1, 2 and QEP3,4 of the Event Manager with Timer2 and Timer 4 as the time base, respectively. Steering servomotor position control is achieved by the length of the pulse method where the width of the control pulse train defines the desired position. PWM5 and PWM6 of TMS320F2812 can be used to generate these control pulses. Timer 3 may be used to generate the servo-control sampling rate.

For ultrasonic ranging, the DSP controller can generate the signal required to trigger pulse input to the sensor causing an ultrasonic burst. The echo pulse triggers the DSP interrupt. An interrupt service routine then measures the time difference between the emission of the burst and reception of the required for time-of-flight calculations. The SHARP IR distance sensors that produce analog outputs can directly be connected to the high-speed ADCs. The bump sensors are generally binary switches which can be connected to the DSP general purpose I/O pin. A receiver and a decoder module can be used for remote control purpose. The impulse signal from the decoder triggers the external interrupt pin of the DSP. The interrupt service routine processes the four digital I/O outputs from the decoder and translates the remote control signal into proper commands. The communication between DSP controller and the host PC is handled by serial communication module

Code Composer Studio (CCS) software development environment allows programming the DSP in C++. On chip software resources allow control of the DC and servo motors by using the built in commands of the control command library. It offers several methods of motor control such as open-loop PWM, closed-loop position control, and closed loop velocity control.

Example 5.3: Highly manoeuvrable small size mobile robot shown in Fig. 5.9 has been developed using a multi-DSP system based on various DSPs from Texas Instrument's C2000 and C5000 families [12]. The steering axles and the driving wheels of this 4-wheel steered robot are driven by six number of geared DC motors. As the power drivers, L293D are used. Six 24-bit counter chips LS7166 are used to process output signals from the six encoders. They are connected to the robot control DSP TMS320LF2407A by a data bus, based on digital I/O. All motors are controlled by PWM signals generated by this control DSP.

An active CCD line camera (SONY ILX551) mounted on the bottom side of robot is used for trajectory following. Analog camera output is digitalized by 8-bit CCD-digitizer MAX1101. The pan-tilt axes of the camera are driven by a DC motor and small servo, respectively. Camera motion control is performed by a second C2000 DSP (TMS320LF2406A), which receives the position commands and other data from the C5500 master DSP through its SPI-port.

Figure 5.9: F.A.A.K. mobile robot system. Courtesy of Control Systems Engineering Group, Fern Universität in Hagen.

The onboard visual system is implemented using the TMS320VC5509A DSP from TI's C5000-family. CCD-digitizer MAX1101 chip communicates with DSP using SPI communication channel. Several IR sensors (SHARP GP2D120) are used for the obstacle detection. The analog outputs of these sensors are converted by ADCs of the control DSP. The robot is equipped with a radio transceiver, Radiometrix BiM2-433, for wireless serial communication with a monitoring system in the host computer that displays the plots or 3D visual robot model-based animation.

Control processor, TMS320LF2407A, performs the following tasks in realtime.

- Robot motion control: Sample and process the sensor signals and computing of the six motor control signals to enable correct and safe navigation of the robot
- Communication with a monitoring system, including visualization of its motion on an host computer
- Execution of challenging intelligent control strategies, e.g. collision-avoidance, self-learning, environment mapping, path planning, etc.

Motion of the active camera is controlled by a second DSP, TMS320LF 2406A, which receives the position commands and other data from the master DSP TMS320VC5509A through its SPI-port. Master DSP performs the following tasks.

- Overall control of system and synchronization of executed tasks
- Image processing and machine vision
- Task planning
- Performing of computationally expensive tasks such as self-learning of motion neuro-fuzzy controller, etc.
- Control of communication with both slave DSPs.

Hence, the multi-DSP system consists of a master DSP (TMS320VC55 09A)and two slave DSPs (TMS320LF2406A and TMS320LF2406A). TMS32 0LF2406A chip also performs additional computational tasks for the master DSP. Because of the high amount of data transfer between between them, a 16KB dual-port SRAM CY7C006AV has been used for communication between TMS320LF2407A and master DSP. Communication between the Master and TMS320LF2406A is based on SPI communication channel.

5.3.5 Active Power Factor Correction

Power factor correction (PFC) in power conversion applications is important for many reasons including to reduce the losses and to improve power quality. PFC often results in significant economic savings. In many applications, an unregulated DC voltage source is obtained by rectifying the AC line voltage and a filter capacitor. Such an unregulated DC voltage is often the input to the DC to DC converters that are widely used in regulated

DC power supplies and DC motor drive applications. However, this results in discontinuous and short duration current spikes being drawn from the mains supply irrespective of the load.

The power factor is the ratio between the real power and apparent power.

$$\text{Power factor} = \frac{\text{Real power [kW]}}{\text{Apparent power [kVA]}}$$

Hence, poor power factor can be as a result of two reasons
- Displacement factor: Phase shift of current with respect to the voltage resulting in displacement
- Distortion factor: Harmonic content present in current resulting in distortion.

Therefore, alternatively

$$\text{Power factor} = \text{Displacement factor} \times \text{Distortion factor}$$
$$= \cos\phi \times \sqrt{\frac{1}{1 + (I_2/I_1)^2 + (I_3/I_1)^2 + \cdots}}$$

where I_1 is the fundamental component and I_i; $i = 2, 3, \cdots$ is the ith current harmonic.

PFC aims at improving the displacement and distortion factors. The PFC strategy is to make the power converter appear as a linear resistor to the supply voltage at low frequencies. This way, the input current waveform can be made to follow the input voltage waveform. Active PFC must control both the input current and the DC bus output voltage. The current is shaped by the rectified line voltage so that the input to the converter appears to be resistive. The output voltage is controlled to reduce the DC bus voltage ripple by changing the average amplitude of the current programming signal.

The dsPIC30F devices contain extensive DSP functionality within a high-performance 16-bit microcontroller architecture [13]. Fig. 5.10 illustrates the digital implementation of PFC using the dsPIC30F6010A for a boost converter. PWM switching pulses are generated by the dsPIC device based on the following three measurements
- Rectified input voltage
- Rectified input current
- DC bus voltage.

The only output from the dsPIC device is firing pulses to the boost converter switch. The output DC voltage of the boost converter and the

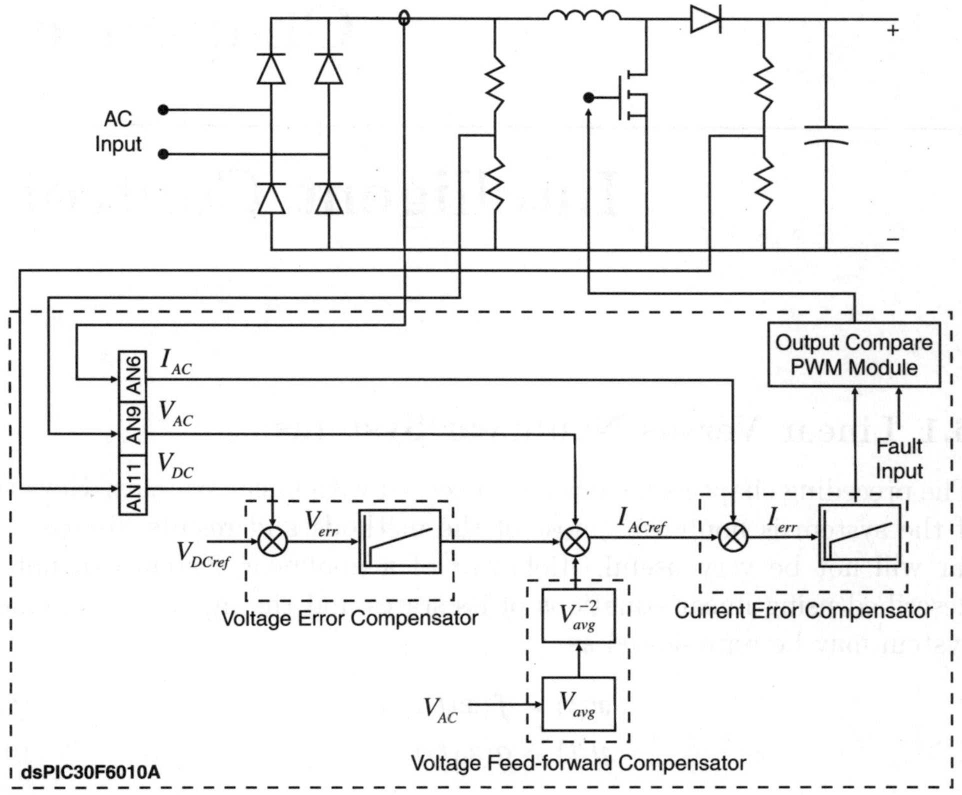

Figure 5.10: Implementation of PFC. Courtesy of Microchip Technologies Inc.

input current through the inductor are the two parameters that are essentially controlled using active PFC. In the technique called the *average current mode control*, the output voltage is controlled by varying the average value of the current amplitude signal. The current signal is calculated digitally by computing the product of the rectified input voltage, the voltage error compensator output and the voltage feed-forward compensator output. The rectified input voltage is multiplied to enable the current signal to have the same shape as the rectified input voltage waveform. The current signal should match the rectified input voltage as closely as possible to have high power factor. The voltage feed-forward compensator is essential for maintaining a constant output power because it compensates for the variations in input voltage from its nominal value [14]. An in-depth theoretical information on this topic may be found in the reference [14].

Chapter 6

Intelligent Control

6.1 Linear Versus Nonlinear Systems

The preceding chapters have been concerned with linear systems. However, if the system is nonlinear, most of the methods and results discussed so far will not be very useful. Behavior of a nonlinear system can not be described using linear equations of its states and the inputs. A nonlinear system may be represented as

$$\dot{\boldsymbol{x}}(t) = \boldsymbol{f}(\boldsymbol{x}(t),\ \boldsymbol{u}(t)) \qquad (6.1)$$
$$\boldsymbol{y}(t) = \boldsymbol{g}(\boldsymbol{x}(t)) \qquad (6.2)$$

where

$$\boldsymbol{u}(t) = [u_1(t)\ u_2(t)\ldots u_p(t)]^T$$

is the input vector and

$$\boldsymbol{y}(t) = [y_1(t)\ y_2(t)\ldots y_q(t)]^T,$$
$$\boldsymbol{x}(t) = [x_1(t)\ x_2(t)\ldots x_n(t)]^T$$

respectively, are the vectors of system outputs and the system states. One can also include the system and measurement noise terms in the above system representation if required. Also note that the linear system representation is a special case of the system representation (6.1) and (6.2).

There are several well-developed techniques for analyzing nonlinear feedback systems. They include describing function method, phase-plane method, Lyapunov-based techniques, and passivity analysis. Special control design techniques for nonlinear systems also exist. They include sliding-mode control, Lyapunov redesign, feedback linearization (e.g., computed-torque control in robotics), gain scheduling, variable structure control, back-stepping, fuzzy-logic control, and neural networks-based control or neurocontrol. It is important to note that as a linear system can be considered as a special case, most of these techniques may be used to solve linear control problems as well.

In contrast, popular methods of the domain of linear systems such as root-locus, Bode plot, Nyquist criterion, state-feedback, pole-placement will be of little or no use when it comes to nonlinear systems. However, if the nonlinearities are significantly smooth, system may be linearized so that linear system techniques that are relatively convenient can be used.

6.2 Linearized Systems

If the nonlinearities are significantly smooth, system may be linearized about the nominal operating point by expanding the nonlinear equations into a Taylor series and dropping out the higher-order terms. When the nonlinearities are due to trigonometric terms, linear models may be arrived at by applying small angle approximation just as how the inverted pendulum equations are linearized to obtain a linearized set of equations in chapter 3.

Consider the nonlinear equation (6.1). Let the nominal operating point and the input be $\bar{\boldsymbol{x}}(t)$ and $\bar{\boldsymbol{u}}(t)$, respectively. Expanding (6.1) about the nominal operating point and neglecting the higher-order terms yield

$$\dot{\boldsymbol{x}} \cong \boldsymbol{f}(\bar{\boldsymbol{x}},\ \bar{\boldsymbol{u}}) + \frac{\partial \boldsymbol{f}}{\partial \boldsymbol{x}}\bigg|_{\bar{\boldsymbol{x}},\bar{\boldsymbol{u}}} \cdot (\boldsymbol{x} - \bar{\boldsymbol{x}}) + \frac{\partial \boldsymbol{f}}{\partial \boldsymbol{u}}\bigg|_{\bar{\boldsymbol{x}},\bar{\boldsymbol{u}}} \cdot (\boldsymbol{u} - \bar{\boldsymbol{u}}) \qquad (6.3)$$

or

$$\dot{x}_i \cong f_i(\bar{\boldsymbol{x}},\ \bar{\boldsymbol{u}}) + \sum_{j=1}^{n}\left[\frac{\partial f_i}{\partial x_j}\bigg|_{\bar{\boldsymbol{x}},\bar{\boldsymbol{u}}} \cdot (x_j - \bar{x}_j)\right] + \sum_{j=1}^{n}\left[\frac{\partial f_i}{\partial u_j}\bigg|_{\bar{\boldsymbol{x}},\bar{\boldsymbol{u}}} \cdot (u_j - \bar{u}_j)\right] \quad (6.4)$$

where $i = 1, 2, \cdots, n$. The preceding equation can be rewritten in the following matrix form.

$$\dot{\boldsymbol{x}} - \dot{\bar{\boldsymbol{x}}} = A(\boldsymbol{x} - \bar{\boldsymbol{x}}) + B(\boldsymbol{u} - \bar{\boldsymbol{u}}) \qquad (6.5)$$

or

$$\Delta\dot{\boldsymbol{x}} = A\,\Delta\boldsymbol{x} + B\,\Delta\boldsymbol{u} \qquad (6.6)$$

where

$$A = \begin{bmatrix} \frac{\partial f_1}{\partial x_1} & \frac{\partial f_1}{\partial x_2} & \cdots & \frac{\partial f_1}{\partial x_n} \\ \frac{\partial f_2}{\partial x_1} & \frac{\partial f_2}{\partial x_2} & \cdots & \frac{\partial f_2}{\partial x_n} \\ \vdots & \vdots & \ddots & \vdots \\ \frac{\partial f_n}{\partial x_1} & \frac{\partial f_n}{\partial x_2} & \cdots & \frac{\partial f_n}{\partial x_n} \end{bmatrix} \qquad (6.7)$$

and

$$B = \begin{bmatrix} \dfrac{\partial f_1}{\partial u_1} & \dfrac{\partial f_1}{\partial u_2} & \cdots & \dfrac{\partial f_1}{\partial u_p} \\[2mm] \dfrac{\partial f_2}{\partial u_1} & \dfrac{\partial f_2}{\partial u_2} & \cdots & \dfrac{\partial f_2}{\partial u_p} \\[2mm] \vdots & \vdots & \ddots & \vdots \\[2mm] \dfrac{\partial f_n}{\partial u_1} & \dfrac{\partial f_n}{\partial u_2} & \cdots & \dfrac{\partial f_n}{\partial u_p} \end{bmatrix} \tag{6.8}$$

are calculated at the nominal operating point, $(\bar{x}, \bar{u})$. The approximate model is linear but valid only for small pertabations about the operating point. In other words, linearity is local.

Example 6.1: A magnetic-ball suspension system consists of a solenoid type electromagnet that is fixed with its axis vertical and a metallic ball. Control objective is to regulate the ball position at a prescribed distance from the end of the electromagnet by regulating the current of the electromagnet. Assume that the supply voltage to the electric circuit is the control input. Also, assume that the center-of-gravity of the ball is right on the vertical axis of the electromagnet. Equation of motion for the vertical linear motion of the ball is

$$m\frac{d^2 x(t)}{dt^2} = mg - k\frac{i^2(t)}{x(t)} \tag{6.9}$$

where $x(t)$ is the vertical distance to the ball from the end of the electromagnet and m is the mass of the ball. For a given electromagnet, k is constant. $i(t)$ is the current in the coil of the electromagnet. $g = 9.81\text{ms}^{-2}$ is the acceleration of gravity. Applying Kirchoff's voltage law to the electrical circuit yields

$$v_{in}(t) = Ri(t) + L\frac{di(t)}{dt} \tag{6.10}$$

where L and R, respectively, are the inductance of the coil and the series resistance of the electrical circuit. $v_{in}(t)$ is the input voltage.

Choose the state variables as

$$\boldsymbol{x}(t) = [x_1(t)\ x_2(t)\ x_3(t)]^T = [x(t)\ \dot{x}(t)\ i(t)]^T. \tag{6.11}$$

The state equations become

$$\dot{x}_1(t) = f_1(\boldsymbol{x}(t),\ u(t)) = x_2(t) \tag{6.12}$$

$$\dot{x}_2(t) = f_2(\boldsymbol{x}(t),\ u(t)) = -\frac{k}{m}\frac{x_3^2(t)}{x_1(t)} + g \tag{6.13}$$

$$\dot{x}_3(t) = f_3(\boldsymbol{x}(t),\ u(t)) = -\frac{R}{L}x_3(t) + \frac{v_{in}(t)}{L} \tag{6.14}$$

which takes the form of (6.1). Linearized model can be obtained as

$$\Delta \dot{\boldsymbol{x}} = A \, \Delta \boldsymbol{x} + B \, \Delta v_{in}(t) \tag{6.15}$$

where A and B are given by

$$A = \begin{bmatrix} \frac{\partial f_1}{\partial x_1} & \frac{\partial f_1}{\partial x_2} & \frac{\partial f_1}{\partial x_3} \\ \frac{\partial f_2}{\partial x_1} & \frac{\partial f_2}{\partial x_2} & \frac{\partial f_2}{\partial x_3} \\ \frac{\partial f_3}{\partial x_1} & \frac{\partial f_3}{\partial x_2} & \frac{\partial f_3}{\partial x_3} \end{bmatrix} = \begin{bmatrix} 0 & 1 & 0 \\ \frac{k}{m}\frac{\bar{x}_3^2}{\bar{x}_1^2} & 0 & -\frac{2k}{m}\frac{\bar{x}_3}{\bar{x}_1} \\ 0 & 0 & -\frac{R}{L} \end{bmatrix} \tag{6.16}$$

and

$$B = \begin{bmatrix} \frac{\partial f_1}{\partial u} \\ \frac{\partial f_2}{\partial u} \\ \frac{\partial f_3}{\partial u} \end{bmatrix} = \begin{bmatrix} 0 \\ 0 \\ \frac{1}{L} \end{bmatrix}. \tag{6.17}$$

In order to evaluate A, the nominal operating point needs to be known. Assume that the ball position is to be regulated at a distance $x = \bar{x} = 0.4$m from the end of the electromagnet. At this equilibrium position, $x_2 = \dot{x} = 0$ and $\dot{x}_2 = 0$. Substituting $\dot{x}_2 = 0$ in (6.14) we get

$$\bar{x}_3 = \sqrt{\frac{m}{k} g \bar{x}_1}. \tag{6.18}$$

If $m = 1$ kg and $k = 1$, $\bar{x}_3 = \sqrt{0.4g}$. Therefore, the nominal operating point is

$$\bar{\boldsymbol{x}} = [\bar{x}_1 \ \bar{x}_2 \ \bar{x}_3]^T = \left[0.4 \ 0 \ \sqrt{0.4g}\right]^T. \tag{6.19}$$

$\bar{u}$ at the equilibrium point can be obtained from (6.14) as $\bar{u} = \bar{v}_{in} = R\bar{x}_3$. However, this is not required in this example. It is now direct to obtain $A|_{(\bar{\boldsymbol{x}},\bar{u})}$ and $B|_{(\bar{\boldsymbol{x}},\bar{u})}$ of the linearized model.

Example 6.2: This example describes obtaining a control law for an active binocular robot head using a linearized kinematics model [28], [29]. A four degrees-of-freedom robotic head tries to track an object to keep it in view performing pure rotations. See Fig. 6.1. Let the head position be parameterized by the vector $\boldsymbol{q}$ which, directly represents head's degrees-of-freedom; pan angle (q_p), tilt angle (q_t), and the vergence angle (q_v), respectively. Let $\boldsymbol{p} = [x_w, y_w, z_w]^T$ be the vector representing the target position in Cartesian coordinates in the world reference frame.

In the control strategy it is a desirable condition that $q_r = -q_l = q_v$, where q_v is the vergence angle as shown in Fig. 6.2. Therefore, the head position vector is $\boldsymbol{q} = [q_p \ q_t \ q_v]^T$.

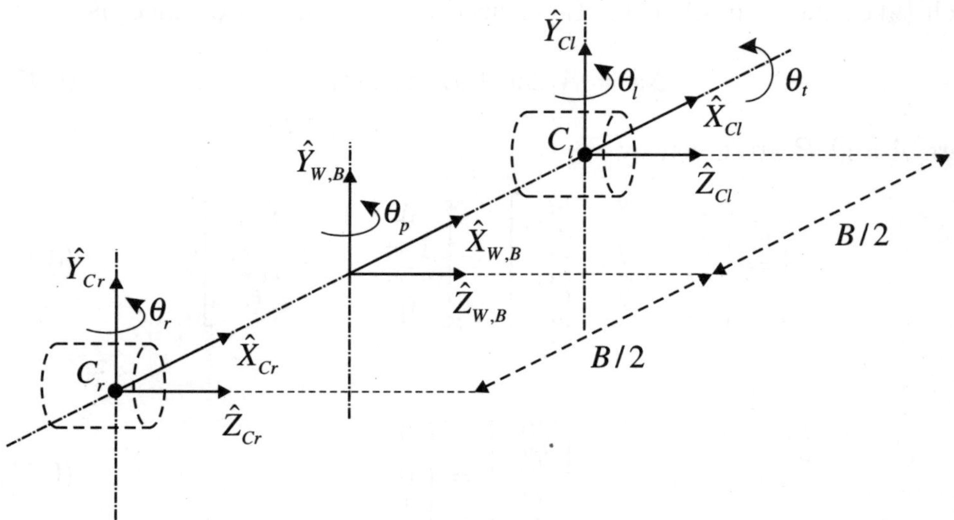

Figure 6.1: Kinematic model of the robotic head.

The visual tracking problem is stated as making a task function defined in image features zero

$$e_{if}(\boldsymbol{f}) = \boldsymbol{f} - \boldsymbol{f}_d \qquad (6.20)$$

where $\boldsymbol{f}$ is the image feature parameter vector defined as follows

$$\boldsymbol{f} = [s_x\ s_y\ d_x]^T = [x_l + x_r\ y_l + y_r\ x_l - x_r]^T \qquad (6.21)$$

where $(x_l, y_l), (x_r, y_r)$ are the image coordinates of the target centroid on left and right cameras.

Since the control task is to track a moving object so as to fixate it at the imaging centers of the two cameras, the desired feature parameter vector $\boldsymbol{f}_d$, understandably, is a three dimensional *null vector*. Since each element of $\boldsymbol{f}$ is a real-valued parameter, we have $\boldsymbol{f} \in \mathcal{F} \subseteq \Re^3$, where $\mathcal{F}$ represents the image feature parameter space.

The mapping from the position of the target to the corresponding image feature parameters can be computed using the projective geometry of the camera, allowing one to write

$$\boldsymbol{F} : {}^C\mathcal{P} \to \mathcal{F}. \qquad (6.22)$$

In (6.22), ${}^C\mathcal{P}$ is the target position space with respect to the camera coordinate frame. See Fig. 6.1 for camera coordinate frames, left camera

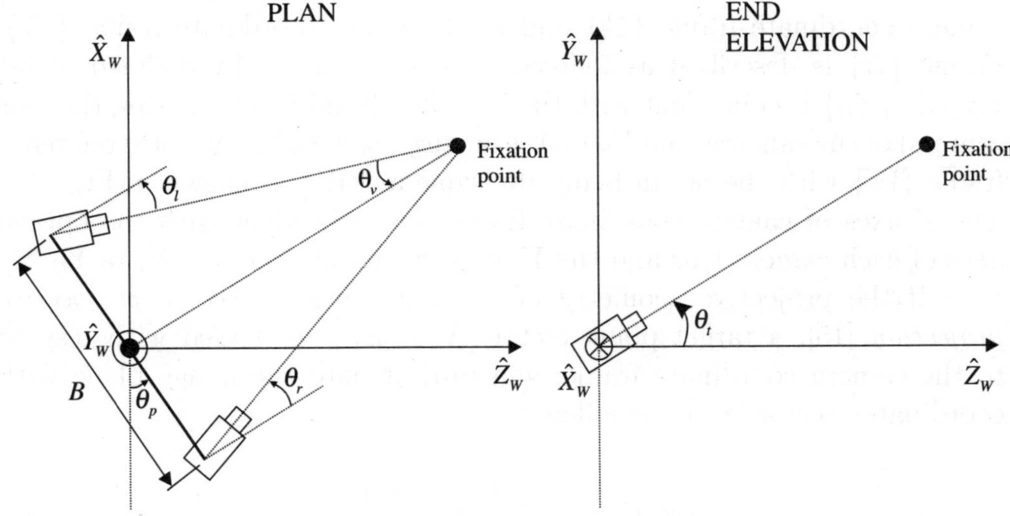

Figure 6.2: Binocular head configuration parameters (q_p, q_t, q_v) with respect to the world coordinate frame $\{\hat{X}_W, \hat{Y}_W, \hat{Z}_W\}$. B is the base-line distance.

coordinate frame $\{C_l\}$ and right camera coordinate frame $\{C_r\}$. Hence, we have

$$f = \mathcal{F}(^{C}\mathcal{P}(q, p)) \tag{6.23}$$

where $^{C}P = {}^{C}\mathcal{P}(q, p)$ is the relative position between target and cameras. When information about the both cameras are stacked together, ^{C}P for this binocular head is given by

$$^{C}P = [^{C_l}P \;\; ^{C_r}P]^T = [X_l \; Y_l \; Z_l \; X_r \; Y_r \; Z_r]^T$$

$$= \begin{bmatrix} (c_p c_l + s_p c_t s_l)x_w + s_t s_l y_w + (-s_p c_l + c_p c_t s_l)z_w - \frac{B}{2}c_l \\ -s_p s_t x_w + c_t y_w - c_p s_t z_w \\ (-c_p s_l + s_p c_t c_l)x_w + s_t c_l y_w + (s_p s_l + c_p c_t c_v)z_w + \frac{B}{2}s_l \\ (c_p c_r - s_p c_t s_r)x_w - s_t s_r y_w - (-s_p c_r + c_p c_t s_r)z_w + \frac{B}{2}c_r \\ -s_p s_t x_w + c_t y_w - c_p s_t z_w \\ (c_p s_r + s_p c_t c_r)x_w + s_t c_r y_w + (-s_p s_r + c_p c_t c_v)z_w + \frac{B}{2}s_r \end{bmatrix}$$

where B is the base-line distance, abbreviations c, s stand for trigonometric functions cos, sin and subscripts p, t, l, r denote the joint angles q_p, q_t, q_l and q_r respectively. This can be obtained using X-Y-Z fixed-angles-method popular in robotics [28]. Coordinate frames involved are: frame $\{B\}$, left camera coordinate frame $\{C_l\}$ and right camera coordinate frame $\{C_r\}$. Frame $\{B\}$ is described as follows: $\hat{X}_B$ of the mutually orthogonal set $\{\hat{X}_B, \hat{Y}_B, \hat{Z}_B\}$ is coincident with the base-line (the line connecting the lens centers) of the cameras and $\hat{Y}_B$ makes an angle q_t with $\hat{Y}_W$ of world reference frame $\{W\}$ with the origin being the same as $\{W\}$ as shown in Fig. 6.1. The Z axes of camera coordinate frames are coincident with the optical axes of each camera lens and the Y axes remain always parallel to $\hat{Y}_B$.

If the projective geometry of a camera is modeled by *perspective projection* [15], a target point vector $[X\ Y\ Z]^T$, expressed with respect to the camera coordinate frame, will project onto the image plane with coordinates vector $[x\ y]^T$ as follows:

$$\pi(X, Y, Z) = \begin{bmatrix} x \\ y \end{bmatrix} = \frac{\lambda}{Z} \begin{bmatrix} X \\ Y \end{bmatrix} \tag{6.24}$$

where λ is the focal length of the camera lens. Stacking information about two cameras together, results in

$$[x_l\ y_l\ x_r\ y_r]^T = \lambda[X_l/Z_l\ Y_l/Z_l\ X_r/Z_r\ Y_r/Z_r]^T \tag{6.25}$$

and hence

$$\boldsymbol{f} = [s_x\ s_y\ d_x]^T = \lambda[X_l/Z_l + X_r/Z_r\ \ Y_l/Z_l + Y_r/Z_r\ \ X_l/Z_l - X_r/Z_r]. \tag{6.26}$$

Expanding (6.23) into a Taylor series about the normal operating point $(\bar{\boldsymbol{q}}, \bar{\boldsymbol{p}})$ and dropping-out higher-order terms yield

$$\boldsymbol{f} \simeq \boldsymbol{f}(\bar{\boldsymbol{q}}, \bar{\boldsymbol{p}}) + \frac{\partial \boldsymbol{f}}{\partial \boldsymbol{q}}\bigg|_{\bar{\boldsymbol{q}}, \bar{\boldsymbol{p}}} \cdot (\boldsymbol{q} - \bar{\boldsymbol{q}}) + \frac{\partial \boldsymbol{f}}{\partial \boldsymbol{p}}\bigg|_{\bar{\boldsymbol{q}}, \bar{\boldsymbol{p}}} \cdot (\boldsymbol{p} - \bar{\boldsymbol{p}}). \tag{6.27}$$

This is a linear kinematics model of the system in the neighborhood of the normal operating conditions. Rearranging (6.27) yields

$$\boldsymbol{f} - \bar{\boldsymbol{f}} = J_q(\bar{\boldsymbol{q}}, \bar{\boldsymbol{p}}) \cdot (\boldsymbol{q} - \bar{\boldsymbol{q}}) \ + \ J_p(\bar{\boldsymbol{q}}, \bar{\boldsymbol{p}}) \cdot (\boldsymbol{p} - \bar{\boldsymbol{p}}) \tag{6.28}$$

where J_q and J_p are the *image Jacobians*, which are evaluated using the following expressions:

$$J_q(\bar{\boldsymbol{q}}, \bar{\boldsymbol{p}}) = \frac{\partial \boldsymbol{f}}{\partial^C \boldsymbol{P}} \cdot \frac{\partial^C \boldsymbol{P}}{\partial \boldsymbol{q}}\bigg|_{\bar{\boldsymbol{q}}, \bar{\boldsymbol{p}}} \ , \quad J_p(\bar{\boldsymbol{q}}, \bar{\boldsymbol{p}}) = \frac{\partial \boldsymbol{f}}{\partial^C \boldsymbol{P}} \cdot \frac{\partial^C \boldsymbol{P}}{\partial \boldsymbol{p}}\bigg|_{\bar{\boldsymbol{q}}, \bar{\boldsymbol{p}}} \tag{6.29}$$

where $^C\boldsymbol{P} = {}^C\boldsymbol{P}(\boldsymbol{q}, \boldsymbol{p})$ and

$$\frac{\partial \boldsymbol{f}}{\partial ^C\!\boldsymbol{P}} = \begin{bmatrix} \frac{\partial s_x}{\partial X_l} & \frac{\partial s_x}{\partial Y_l} & \frac{\partial s_x}{\partial Z_l} & \frac{\partial s_x}{\partial X_r} & \frac{\partial s_x}{\partial Y_r} & \frac{\partial s_x}{\partial Z_r} \\[4pt] \frac{\partial s_y}{\partial X_l} & \frac{\partial s_y}{\partial Y_l} & \frac{\partial s_y}{\partial Z_l} & \frac{\partial s_y}{\partial X_r} & \frac{\partial s_y}{\partial Y_r} & \frac{\partial s_y}{\partial Z_r} \\[4pt] \frac{\partial d_x}{\partial X_l} & \frac{\partial d_x}{\partial Y_l} & \frac{\partial d_x}{\partial Z_l} & \frac{\partial d_x}{\partial X_r} & \frac{\partial d_x}{\partial Y_r} & \frac{\partial d_x}{\partial Z_r} \end{bmatrix},$$

$$\frac{\partial ^C\!\boldsymbol{P}}{\partial \boldsymbol{q}} = \begin{bmatrix} \frac{\partial X_l}{\partial q_p} & \frac{\partial X_l}{\partial q_t} & \frac{\partial X_l}{\partial q_v} \\[4pt] \frac{\partial Y_l}{\partial q_p} & \frac{\partial Y_l}{\partial q_t} & \frac{\partial Y_l}{\partial q_v} \\[4pt] \frac{\partial Z_l}{\partial q_p} & \frac{\partial Z_l}{\partial q_t} & \frac{\partial Z_l}{\partial q_v} \\[4pt] \frac{\partial X_r}{\partial q_p} & \frac{\partial X_r}{\partial q_t} & \frac{\partial X_r}{\partial q_v} \\[4pt] \frac{\partial Y_r}{\partial q_p} & \frac{\partial Y_r}{\partial q_t} & \frac{\partial Y_r}{\partial q_v} \\[4pt] \frac{\partial Z_r}{\partial q_p} & \frac{\partial Z_r}{\partial q_t} & \frac{\partial Z_r}{\partial q_v} \end{bmatrix}, \qquad \frac{\partial ^C\!\boldsymbol{P}}{\partial \boldsymbol{p}} = \begin{bmatrix} \frac{\partial X_l}{\partial x_W} & \frac{\partial X_l}{\partial y_W} & \frac{\partial X_l}{\partial z_W} \\[4pt] \frac{\partial Y_l}{\partial x_W} & \frac{\partial Y_l}{\partial y_W} & \frac{\partial Y_l}{\partial z_W} \\[4pt] \frac{\partial Z_l}{\partial x_W} & \frac{\partial Z_l}{\partial y_W} & \frac{\partial Z_l}{\partial z_W} \\[4pt] \frac{\partial X_r}{\partial x_W} & \frac{\partial X_r}{\partial y_W} & \frac{\partial X_r}{\partial z_W} \\[4pt] \frac{\partial Y_r}{\partial x_W} & \frac{\partial Y_r}{\partial y_W} & \frac{\partial Y_r}{\partial z_W} \\[4pt] \frac{\partial Z_r}{\partial x_W} & \frac{\partial Z_r}{\partial y_W} & \frac{\partial Z_r}{\partial z_W} \end{bmatrix}.$$

The elements q_p and q_t of $\boldsymbol{q}$ can directly be obtained from the head's internal encoders and the vergence angle is approximated by $q_v = (|q_l| + |q_r|)/2$. When the target motion parameters and their variations are known, a suitable kinematic expression to compute desired head position so as to compensate for feature variations is given by

$$\boldsymbol{q} - \bar{\boldsymbol{q}} = J_q^{-1} \cdot (\boldsymbol{f} - \bar{\boldsymbol{f}}) - J_q^{-1} J_p \cdot (\boldsymbol{p} - \bar{\boldsymbol{p}}). \tag{6.30}$$

When (6.28) is rewritten using discrete notations, we have

$$\boldsymbol{f}(k+1) - \boldsymbol{f}(k) = J_{q_s}(k)[\boldsymbol{q}(k+1) - \boldsymbol{q}(k)] + J_p(k)[\boldsymbol{p}(k) - \boldsymbol{p}(k-1)] \tag{6.31}$$

where k denotes the time index. Note that since no target motion predictions are made, immediate future changes in target configuration parameters are approximated by their backward-difference. Modifying (6.30) as

$$\boldsymbol{q}(k+1) - \boldsymbol{q}(k) = J_q^{-1}(k)\{\boldsymbol{f}_d(k+1) - \boldsymbol{f}(k) - J_p(k)[\boldsymbol{p}(k) - \boldsymbol{p}(k-1)]$$
$$+ K[\boldsymbol{f}_d(k) - \boldsymbol{f}(k)]\} \tag{6.32}$$

and combining with (6.31) gives a new task function given by

$$[\boldsymbol{f}_d(k+1) - \boldsymbol{f}(k+1)] + K[\boldsymbol{f}_d(k) - \boldsymbol{f}(k)] = \boldsymbol{0} \tag{6.33}$$

or

$$\boldsymbol{e}_{if}(k+1) + K\boldsymbol{e}_{if}(k) = \boldsymbol{0} \qquad (k = 0, 1, 2, \ldots) \tag{6.34}$$

where $K \in \Re^{3 \times 3}$ is a diagonal constant gain matrix with its elements $-1 \leq k_j \leq 1$, $(j = 1, \ldots, n)$. This allows us to write

$$\boldsymbol{f}(k + \ell) = (-1)^{\ell} K^{\ell}\, \boldsymbol{f}(k) \qquad (\ell = 1, 2, 3, \ldots). \tag{6.35}$$

Therefore, feature parameters will always be converged to zero. Note that, any finite-difference approximation to the partial derivatives is not made in the difference equation representation in (6.32). They are computed using (6.29) at each operating point $t = kT$, where T is the sampling interval. Understandably, $\boldsymbol{q}_s(k + 1)$ is the desired head configuration parameter vector $[q_{pd}\ q_{td}\ q_{vd}]^T$. The desired head trajectory is $\boldsymbol{q}_d = [q_{pd}\ q_{td}\ q_{ld}\ q_{rd}]^T$, with $-q_{ld} = q_{rd} = q_{vd}$.

In this tracking control problem, change in target position should be known. However, $(\boldsymbol{p}(k) - \boldsymbol{p}(k - 1))$ may be assumed to be negligible if the target motion is slow when compared to the frame rate of the vision system. This greatly simplifies (6.32).

The linearized kinematics model derived in this example may be used to implement kinematics-based visual tracking control. Moreover, the desired head trajectories generated by the use of this model may be used as the reference points to a high-end tracking control algorithm to implement more advanced control as in [29].

6.3 Lyapunov-based Stability Analysis of Systems

Lyapunov-based methods are the most general methods for the determination of stability of nonlinear and/or time-varying systems. The method applies to system of any order. Nevertheless, finding a suitable Lyapunov function (energy-like function as will be discussed shortly) may be quite difficult and the mathematical analysis may also look comparatively complex. The mathematical foundation required for Lyapunov-based stability analysis of dynamical systems is laid first.

6.3.1 Mathematical Background

Following definitions are required before proceeding to the Lyapunov method of stability analysis examples given in this chapter.

6.3.1.1 Definiteness of a scalar function: A scalar function $L(\boldsymbol{x})$ is said to be

1. positive definite (PD) in a region that includes the origin of the state space if $L(\boldsymbol{x}) > 0$ for all states, $\boldsymbol{x}$, in that region except at the origin where $L(\boldsymbol{0}) = 0$.

2. negative definite if $-L(\boldsymbol{x})$ is PD.

3. positive semi-definite in a region that includes the origin of the state space if it is positive at all states in that region except at the origin and at certain other states where it is zero.

4. negative semi-definite if $-L(\boldsymbol{x})$ is positive semi-definite.

6.3.1.2 Definiteness of a matrix: A matrix $P = [p_{ij}] \in \Re^{n \times n}$ is said to be positive definite (PD) if all the principal minors of P are positive, i.e.,

$$p_{11} > 0, \quad \begin{vmatrix} p_{11} & p_{12} \\ p_{21} & p_{21} \end{vmatrix} > 0, \cdots, \begin{vmatrix} p_{11} & p_{12} & \cdots & p_{1n} \\ p_{21} & p_{22} & \cdots & p_{2n} \\ \vdots & \vdots & \ddots & \vdots \\ p_{n1} & p_{n2} & \cdots & p_{nn} \end{vmatrix} > 0.$$

Note also that $-P$ is negative definite.

A class of scalar functions of the following form plays and important role in Lyapunov-based stability analysis. An example is

$$L(\boldsymbol{x}) = \boldsymbol{x}^T P \boldsymbol{x} \tag{6.36}$$

where $\boldsymbol{x}$ is a real vector. Simple computations can prove that $L(\boldsymbol{x})$ is PD if P is real, symmetric, and PD. Note in the sequel of this chapter that the mass-inertia matrix of rigid robot systems is a real, symmetric, and PD. We will also see that this matrix from the system dynamics can be used to realize PD Lyapunov functions in the stability proofs.

6.3.1.3 Signum function of a matrix   Some mathematical preliminaries that will be used in the later chapters are stated here. Let $\Re$ denote the real scalars, $\Re^n$ denote the real n-vectors, and $\Re^{m \times n}$ denote the real $m \times n$-matrices. Signum function of a matrix, $A = [a_{i,j}]_{m \times n} \in \Re^{m \times n}$, is defined as

$$\mathrm{sgn}(A) = [\mathrm{sgn}(a_{i,j})]_{m \times n}$$

where

$$\mathrm{sgn}(a_{i,j}) = \begin{cases} 1 & \text{if } a_{i,j} > 0 \\ 0 & \text{if } a_{i,j} = 0 \\ -1 & \text{if } a_{i,j} < 0 \end{cases}$$

6.3.1.4 Norm of a matrix   Norm of a matrix vector, $\boldsymbol{x} \in \Re^n$, is defined by

$$||\boldsymbol{x}||^2 = \boldsymbol{x}^T \boldsymbol{x}.$$

Given $A = [a_{ij}] \in \Re^{m \times n}$ the Frobenius norm of a matrix is defined by

$$\|A\|_F^2 = \operatorname{tr}(A^T A) = \sum_{i,j} a_{ij}^2$$

and the Euclidean norm by

$$\|A\|^2 = \lambda_{\max}(A^T A)$$

where $\lambda_{\max}(\cdot)$ is the maximum eigenvalue of $(\cdot)$ and $\operatorname{tr}(\cdot)$ is the trace. The trace of a matrix is the sum of the main diagonal elements of that matrix.

6.3.1.5 β-ball lemma   A modified version of the β-ball lemma [16], [17] will be needed in the sequel. Consider a dynamical system

$$\dot{\boldsymbol{x}}_i = \boldsymbol{f}_i(\boldsymbol{x}_1, \cdots, \boldsymbol{x}_m), \quad \boldsymbol{x}_i \in \Re^n, \quad t \geq 0, \quad i = 1, \cdots, m.$$

Let $\boldsymbol{f}_i(\cdot)$ be locally Lipschitz with respect to $\boldsymbol{x}_1, \cdots, \boldsymbol{x}_m$. Suppose a function $V(\cdot) : \Re^{n \times m} \to \Re^+$ is given such that

$$V(\boldsymbol{x}_1, \cdots, \boldsymbol{x}_m) = \sum_{i,j=1}^m \boldsymbol{x}_i^T P_{ij} \boldsymbol{x}_j$$

where for each $i = 1, \cdots, m$ there exists a $\xi_i > 0$ such that

$$\xi_i \|\boldsymbol{x}_i\|^2 \leq V(\boldsymbol{x}_1, \cdots, \boldsymbol{x}_m)$$

$$\dot{V}(\boldsymbol{x}_1, \cdots, \boldsymbol{x}_m) = -\sum_{i \in I_1} \left(\alpha_i - \sum_{j \in I_2} \gamma_{ij} \|\boldsymbol{x}_j\| \right) \|\boldsymbol{x}_i\|^2$$

where $\alpha_i, \gamma_{ij} > 0$, $I_2 \subset I_1 \subset \{1, \cdots, m\}$. Define $V_0 \equiv V(\boldsymbol{x}_1(0), \cdots, \boldsymbol{x}_m(0))$. If for all $i \in I_1$

$$\bar{\alpha}_i \equiv \alpha_i - \sum_{j \in I_2} \gamma_{ij} V_0^{\frac{1}{2}} \xi_j^{-\frac{1}{2}} > 0$$

then for all $\kappa_i \in [0, \bar{\alpha}_i]$, the following inequality holds

$$\dot{V}(\boldsymbol{x}_1, \cdots, \boldsymbol{x}_m) \leq -\sum_{i \in I_1} \kappa_i \|\boldsymbol{x}_i\|^2, \quad \text{for all } t \geq 0.$$

6.3.2 Stability in the Sense of Lyapunov

Consider the nonlinear system

$$\dot{x} = f(x) \qquad (6.37)$$

where $x = [x_1 \ x_2, \cdots, \ x_n]^T$ is the vector of states. Assume that the system (6.37) has an unique solution, $x = \phi(t; x_0, t_0)$, for the given initial condition, $x = x_0$ at $t = t_0$. Assume that the origin is the only equilibrium state. That is

$$f(0) = 0 \quad \text{for all } t. \qquad (6.38)$$

6.3.2.1 Stability in the sense of Lyapunov: Let $S(\delta)$ consists of all the points such that

$$\sqrt{x_{10}^2 + x_{20}^2 + \cdots + x_{n0}^2} = ||x_0(t)|| \leq \delta \qquad (6.39)$$

which is a hyper-spherical region centered at the origin and defines the bounds of the original state or the starting point, x_0. Likewise, let $S(\epsilon)$ consists of all points such that

$$||\phi(t; x_0, t_0)|| \leq \epsilon. \qquad (6.40)$$

The equilibrium state (here, the origin) of the system (6.37) is said to be stable in the sense of Lyapunov if, there exists an $S(\delta)$ such that trajectories starting in $S(\delta)$ do not leave $S(\epsilon)$ as $t \to \infty$.

If δ does not depend on t_0, the equilibrium state is said to be uniformly stable. The equilibrium state is said to be asymptotically stable if every trajectory starting within $S(\delta)$ converges to the equilibrium state as $t \to \infty$.

6.3.2.2 Lyapunov method of stability analysis: The method is stated as a theorem as follows.

Theorem: Consider the nonlinear system, $\dot{x} = f(x)$ wher $f(0) = 0$ for all t. If there exists a function, $L(x)$, that is positive definite with the first time derivative, $\dot{L}(x)$, that is negative definite, then the equilibrium state at the origin is uniformly asymptotically stable.

If $L(x)$ is positive definite and $\dot{L}(x)$ is negative semi-definite, then the system can remain in a limit cycle. The equilibrium state at the origin, in this case, is said to be stable in the sense of Lyapunov.

Lypunov function is an energy-like function. For instance, consider a vibrating system. Understandably, total kinetic energy (a PD function) should continually decrease for the system to reach the equilibrium state where velocity is zero. When this happens, the first time derivative of total energy must be negative definite.

6.3.2.3 UUB stability: Consider the nonlinear system

$$\dot{x} = f(x, t)$$

where x is the vector of states. We say the solution is uniformly ultimately bounded (UUB) if there exists a compact set U such that, for all $x(t_0) = x_0 \in U$, there exists an $\varepsilon > 0$ and a number $T(\varepsilon, x_0)$ such that $||x(t)|| < \varepsilon$ for all $t \geq t_0 + T$.

Note that this stability notion that is widely used in robotics literature is only a variant of the Lyapunov method.

6.3.3 Stability Analysis of Linear Systems

Consider the linear time-invariant version of (6.37) given by

$$\dot{x} = Ax \tag{6.41}$$

where $A \in \Re^{n \times n}$ is a constant non-singular matrix. Note that the only equilibrium state is the origin, $x = 0$.

Consider the following Lyapunov function candidate

$$L(x) = x^T P x \tag{6.42}$$

where P is a positive definite (PD) real symmetric matrix. Note that $L(x) > 0$ for $x \neq 0$. Taking the time derivative of the Lyapunov function yields

$$\begin{aligned}
\dot{L}(x) &= \dot{x}^T P x + x^T P \dot{x} \\
&= (Ax)^T P x + x^T P A x \\
&= x^T A^T P x + x^T P A x \\
&= x^T (A^T P + P A) x.
\end{aligned}$$

For $\dot{L}(x)$ to be negative definite, which is the requirement for asymptotic stability

$$Q = -(A^T P + P A) \tag{6.43}$$

must be positive definite. Thus, a necessary and sufficient condition for asymptotic stability of a linear system, $\dot{x} = Ax$, is that there must exist a PD symmetric matrix, P, such that $Q = -(A^T P + P A)$ is PD. It can also be shown that $L(x) \to \infty$ as $||x|| \to \infty$ and hence the origin is asymptotically stable in the large. Proof of this result is omitted in this book.

Example 6.3: Consider the 2nd order system described by

$$\begin{bmatrix} \dot{x}_1 \\ \dot{x}_2 \end{bmatrix} = \begin{bmatrix} -1 & -2 \\ 1 & -4 \end{bmatrix} \begin{bmatrix} x_1 \\ x_2 \end{bmatrix}.$$

Clearly, origin is the only equilibrium state. To determine the stability of this state, we may find whether a PD real symmetric matrix, P, exists so that $Q = -(A^T P + PA)$ is PD. One may choose the simplest PD matrix $Q = I$, the identity matrix. Now, the following equation can be solved to obtain P.

$$A^T P + PA = -I$$

$$\begin{bmatrix} -1 & 1 \\ -2 & -4 \end{bmatrix} \begin{bmatrix} p_{11} & p_{12} \\ p_{12} & p_{22} \end{bmatrix} + \begin{bmatrix} p_{11} & p_{12} \\ p_{12} & p_{22} \end{bmatrix} \begin{bmatrix} -1 & -2 \\ 1 & -4 \end{bmatrix} = \begin{bmatrix} -1 & 0 \\ 0 & -1 \end{bmatrix}.$$

This yields the following linear equations

$$-2p_{11} + 2p_{12} = -1$$
$$-2p_{11} - 5p_{12} + p_{22} = 0$$
$$-4p_{12} - 8p_{22} = -1$$

whence solved, we get

$$P = \frac{1}{60} \begin{bmatrix} 23 & -7 \\ -7 & 11 \end{bmatrix}.$$

Check the determinants of the successive principal minors

$$23 > 0, \quad \begin{vmatrix} 23 & -7 \\ -7 & 11 \end{vmatrix} > 0.$$

Clearly, P is PD. Hence the equilibrium state at the origin is asymptotically stable in the large. The Lyapunov function is

$$L(x) = x^T P x = \frac{1}{60}(23x_1^2 - 14x_1 x_2 + 11x_2^2) \tag{6.44}$$

and

$$\dot{L}(x) = -\frac{1}{60}(32x_1^2 + 16x_2^2). \tag{6.45}$$

6.4 Robot Control

The robot system shown in Fig. 6.3 by Neuronics AG is typically used for various automation applications such as handling, assembly, production, and laboratory automation.

Increasing demands on the performance of robot manipulators in terms of accuracy and speed and the wide range of applications including heavy usage in the industry has led to development of various control techniques. Rigid robot systems such as industrial robot manipulators are a good practical example for complex nonlinear systems. Design and analysis of control algorithms for such systems is often based on Lyapunov methods.

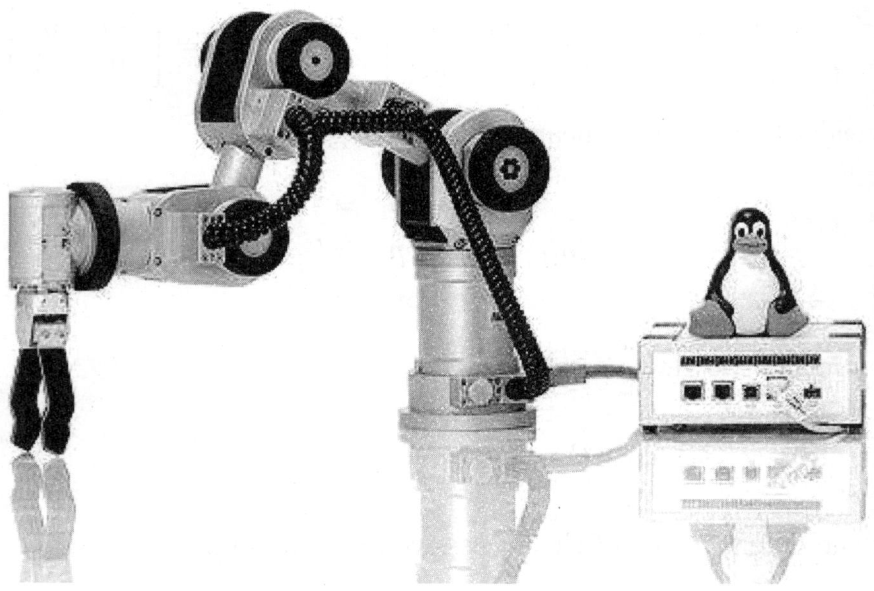

Figure 6.3: The Katana robotic arm. Courtesy of Neuronics AG.

6.4.0.1 Robot Dynamics   In the presence of disturbances and friction, the dynamic model of an n degrees-of-freedom rigid robot system (*eg: n*-link rigid robot manipulator) may be expressed as [36]

$$\tau = M(\boldsymbol{q})\ddot{\boldsymbol{q}} + C(\boldsymbol{q}, \dot{\boldsymbol{q}})\dot{\boldsymbol{q}} + G(\boldsymbol{q}) + F_v\dot{\boldsymbol{q}} + F_s(\boldsymbol{q}) \tag{6.46}$$

where $\boldsymbol{q} \in \Re^n$ the joint variable vector, $M(\boldsymbol{q})$ is the inertia matrix, $C(\boldsymbol{q}, \dot{\boldsymbol{q}})$ the coriolis/centripetal matrix, $G(\boldsymbol{q})$ the gravity vector, $F_v\dot{\boldsymbol{q}}$ is the viscous friction wherein $F_v \in \Re^{n \times n} > 0$ is constant with $F_{v,m} \leq ||F_v|| \leq F_{v,M}$. $F_s(q) \in \Re^n$ is assumed to be a continuous function of q.

For a planar (vertical) two-link robot arm, for instance, the joint variable is $q = [q_1 \; q_2]^T$. The dynamics of such a robot arm are given below [35].

$$M(q) = \begin{bmatrix} 9.77 + 2.02\cos(q_2) & 1.26 + 1.01\cos(q_2) \\ 1.26 + 1.01\cos(q_2) & 1.12 \end{bmatrix}$$

$$C(q, \dot{q}) = \begin{bmatrix} -1.01\sin(q_2)\dot{q}_2 & -1.01\sin(q_2)(\dot{q}_1 + \dot{q}_2) \\ 1.01\sin(q_2)\dot{q}_1 & 0 \end{bmatrix}$$

$$G(q) = g \begin{bmatrix} 8.1\sin(q_1) + 1.13\sin(q_1 + q_2) \\ 1.13\sin(q_1 + q_2) \end{bmatrix}$$

where $g = 9.8\text{ms}^{-2}$ is the acceleration of gravity. The friction terms may be assumed to be

$$F_v = [1.0 \; 1.2]^T, \quad F_s(q) = [10\cos(3q_1) \; 5\sin(2q_2)]^T.$$

Property 1: Mathematical models of certain systems have unique properties that can be used in the controller design and analysis. Rigid revolute robot systems are a good example for this. For robot systems with only revolute joints, the matrices $M(q)$ and $C(q, \dot{q})$ are bounded by

$$||M^{-1}(q)|| \leq M_M, \quad ||C(q, x)|| \leq C_M ||x|| \quad \forall \, q, x \tag{6.47}$$

and satisfies

$$C(q, x)y = C(q, y)x \quad \forall \, x, y. \tag{6.48}$$

These properties will be used in the robot controller design and stability analysis examples given in the sequel of this chapter.

6.4.1 Feedback Linearizing Control

Among the various control techniques that have been developed for industrial robot control, computed-torque-control (CTC), arguably, is the most well known. Lagrangian dynamics of the robot manipulator allows one to write dynamics equations in terms of joint motor torques as in (6.46). The computed torque controller feedback linearizes the nonlinear system using exact knowledge of the system dynamics. To that end, the major drawbacks of CTC are it requires an accurate knowledge of the dynamics that is a strong requirement in practice and it requires measurement of joint angle velocities (in addition to the positions) that increases the cost of implementation and the weight of the robot system. Nevertheless, simplicity and guaranteed exponential stability have made CTC so popular.

CTC controllers can be designed for other robotic systems such as robot heads as well. The aforementioned drawbacks may be overcome, for instance, by designing velocity observers and using neural networks to learn complex model dynamics as described in the following sections of this chapter.

The CTC controller that is often used in order to achieve tracking control in robot manipulators is given by

$$\begin{cases} \boldsymbol{\tau}(t) &= \hat{M}(\boldsymbol{q})\boldsymbol{\nu} + \hat{C}(\boldsymbol{q}, \dot{\boldsymbol{q}})\dot{\boldsymbol{q}} + \hat{G}(\boldsymbol{q}) + \hat{F}_v\dot{\boldsymbol{q}} + \hat{F}_s(\boldsymbol{q}) \\ \boldsymbol{\nu} &= \ddot{\boldsymbol{q}}_d - K_d\dot{\boldsymbol{e}} - K_p\boldsymbol{e} \end{cases} \tag{6.49}$$

where $\boldsymbol{q} \in \Re^n$ is the joint variable vector of an n-degrees-of-freedom robot system and $\boldsymbol{e} = \boldsymbol{q}_d - \boldsymbol{q}$ is the position error vector between the desired position and the sensed position, and $K_d, K_p \in \Re^{n \times n}$ are diagonal gain matrices with constant positive elements k_j; $j = 1, 2, \cdots, n$. $\hat{(\cdot)}$ is the estimated value of $(\cdot)$. If $\hat{(\cdot)} = (\cdot)$, then the robot system (6.46) with the controller (6.49) results in the error dynamics

$$\ddot{\boldsymbol{e}} + K_d\dot{\boldsymbol{e}} + K_p\boldsymbol{e} = 0 \tag{6.50}$$

that shows that the exponential stability of the error can be guaranteed by suitably selecting the gains, K_d and K_p. However, if the system dynamics are not known sufficiently accurately, controller performance degrades even leading to instability. Accurate modeling of complex robot systems is practically difficult as friction terms, in particular, are too complex to be modeled and bound to vary over time. To overcome this problem, one may design and insert a neural networks compensator to counteract for modeled and unmodeled uncertainties. One such example will be described in the following sections.

6.4.2 Nonlinear Controller-observer Schemes

A controller-velocity observer scheme published in [16] is discussed first. Here also, it is assumed that the robot dynamics are accurately known. In the section that follows, this problem is overcome by including an online neural estimator to learn the robot system dynamics hence eliminating the requirement of an exact dynamic model.

In the sequel, $\hat{(\cdot)}$ denotes the estimated value of $(\cdot)$. Define $\boldsymbol{e} = \boldsymbol{q} - \boldsymbol{q}_d$, and $\tilde{\boldsymbol{q}} = \boldsymbol{q} - \hat{\boldsymbol{q}}$ where $\boldsymbol{q}_d \in \Re^n$ is the desired joint angle trajectory.

It is assumed that $\boldsymbol{q}_d$ and its time derivatives, $\dot{\boldsymbol{q}}_d$, $\ddot{\boldsymbol{q}}_d$, are known. The control objective is to regulate the tracking error, $\boldsymbol{e}$. Implicit in it are simultaneously keeping the observer estimation error small.

Consider the following combined controller-velocity observer system:

Controller:

$$\begin{cases} \boldsymbol{\tau}(t) &= M(\boldsymbol{q})\boldsymbol{\nu} + C(\boldsymbol{q},\dot{\boldsymbol{q}}_d)\dot{\hat{\boldsymbol{q}}} + G(\boldsymbol{q}) \\ \boldsymbol{\nu} &= \ddot{\boldsymbol{q}}_d - K_d(\dot{\hat{\boldsymbol{q}}} - \dot{\boldsymbol{q}}_d) - K_p(\boldsymbol{q} - \boldsymbol{q}_d) \end{cases} \tag{6.51}$$

Observer:

$$\begin{cases} \dot{\hat{\boldsymbol{q}}} &= \boldsymbol{z} + L_d(\boldsymbol{q} - \hat{\boldsymbol{q}}) \\ \dot{\boldsymbol{z}} &= \boldsymbol{\nu} + L_p(\boldsymbol{q} - \hat{\boldsymbol{q}}) \end{cases} \tag{6.52}$$

Matrices $K_d > 0$, $K_p > 0$ are assumed to be diagonal. $\dot{\boldsymbol{z}} \in \Re^n$ denotes a reference acceleration input, which is obtained by modifying the resolved acceleration with position estimation error. Integrating $\dot{\boldsymbol{z}}$ and further modifying it by position estimation error yield the estimated velocity.

Consider the controller-observer system defined by (6.51) and (6.52) in a closed loop with the robot system (6.46). The closed-loop system can be made to be UUB by suitably selecting the controller-observer gains, K_p, K_d, L_p, and L_d.

Proof: In the sequel, we denote the minimum and maximum eigenvalue of any matrix $A(\boldsymbol{x}) = A^T(\boldsymbol{x}) > 0$ for all $\boldsymbol{x} \in \Re^n$ by A_m and A_M. Consider the Lyapunov function candidate

$$L = \frac{1}{2}\dot{\tilde{\boldsymbol{q}}}^T\dot{\tilde{\boldsymbol{q}}} + \lambda\dot{\tilde{\boldsymbol{q}}}^T\tilde{\boldsymbol{q}} + \frac{1}{2}\tilde{\boldsymbol{q}}^T(L_p + \lambda L_d)\tilde{\boldsymbol{q}} + \frac{1}{2}\dot{\boldsymbol{e}}^T\dot{\boldsymbol{e}} + \lambda\dot{\boldsymbol{e}}^T\boldsymbol{e}$$
$$+ \frac{1}{2}\boldsymbol{e}^T(K_p + \lambda K_d)\boldsymbol{e} \tag{6.53}$$

which is positive definite for λ sufficiently small where $\Lambda = \Lambda^T > 0$ is a constant matrix and $\text{tr}(\cdot)$ is the trace of matrix $(\cdot)$. With (6.51), (6.52), and (6.46), we obtain

$$M(\boldsymbol{q})(\ddot{\tilde{\boldsymbol{q}}} + L_d\dot{\tilde{\boldsymbol{q}}} + L_p\tilde{\boldsymbol{q}}) = C(\boldsymbol{q},\dot{\boldsymbol{q}}_d)\dot{\hat{\boldsymbol{q}}} - C(\boldsymbol{q},\dot{\boldsymbol{q}})\dot{\boldsymbol{q}} - F_v\dot{\boldsymbol{e}} \tag{6.54}$$

$$M(\boldsymbol{q})(\ddot{\boldsymbol{e}} + K_d\dot{\boldsymbol{e}} + K_p\boldsymbol{e} - K_d\dot{\tilde{\boldsymbol{q}}}) = C(\boldsymbol{q},\dot{\boldsymbol{q}}_d)\dot{\hat{\boldsymbol{q}}} - C(\boldsymbol{q},\dot{\boldsymbol{q}})\dot{\boldsymbol{q}} - F_v\dot{\boldsymbol{e}}. \tag{6.55}$$

Time differentiating (6.53), yields

$$\dot{L} = -\dot{\tilde{\boldsymbol{q}}}^T(L_d - \lambda I)\dot{\tilde{\boldsymbol{q}}} - \lambda\tilde{\boldsymbol{q}}^T L_p\tilde{\boldsymbol{q}} - \dot{\boldsymbol{e}}^T(K_d - \lambda I)\dot{\boldsymbol{e}}$$
$$-\lambda\boldsymbol{e}^T K_p\boldsymbol{e} + (\dot{\tilde{\boldsymbol{q}}} + \lambda\tilde{\boldsymbol{q}} + \dot{\boldsymbol{e}} + \lambda\boldsymbol{e})^T M^{-1}(\boldsymbol{q})$$
$$\times[C(\boldsymbol{q},\dot{\boldsymbol{q}}_d)\dot{\boldsymbol{q}}_d - C(\boldsymbol{q},\dot{\boldsymbol{q}})\dot{\boldsymbol{q}} - F_v\dot{\boldsymbol{e}}] + (\dot{\boldsymbol{e}} + \lambda\boldsymbol{e})^T K_d\dot{\tilde{\boldsymbol{q}}}. \tag{6.56}$$

It is shown that (6.56) can be upper bounded by

$$\dot{L} \leq -\alpha_2||e||^2 - \alpha_3||\dot{\tilde{q}}||^2 - \alpha_4||\tilde{q}||^2$$
$$-(\alpha_1 - \gamma_1||\dot{\tilde{q}}|| - \gamma_2||\tilde{q}|| - \gamma_3||\dot{e}|| - \gamma_4||e||)||\dot{e}||^2 \qquad (6.57)$$

where

$$\alpha_1 = (K_{d,m} - \lambda - 5\beta - \frac{1}{2}\varrho^2 K_{d,M})$$

$$\alpha_2 = (\lambda K_{p,m} - \lambda^2\beta - \frac{1}{2}\lambda^2 K_{d,M})$$

$$\alpha_3 = (L_{d,m} - \lambda - \beta - \frac{1}{2}(1 - \varrho^{-2})K_{d,M} - 2M_M)$$

$$\alpha_4 = (\lambda_{p,m} - \lambda^2\beta)$$

$$\lambda\gamma_1 = \gamma_2 = \lambda\gamma_3 = \gamma_4 = \lambda M_M C_M. \qquad (6.58)$$

Proof: Define $\beta = ||\dot{q}_d||M_M C_M + \frac{1}{2}M_M F_{v,M}$. Following inequalities can be written for the terms of (6.56)

$$-\dot{\tilde{q}}^T(L_d - \lambda I)\dot{\tilde{q}} - \lambda\tilde{q}^T L_p \tilde{q} - \dot{e}^T(K_d - \lambda I)\dot{e} - \lambda e^T K_p e$$
$$\leq -(L_{d,m} - \lambda)||\dot{\tilde{q}}||^2 - \lambda L_{p,m}||\tilde{q}||^2 - (K_{d,m} - \lambda)||\dot{e}||^2$$
$$-\lambda K_{p,m}||e||^2 \qquad (6.59)$$

$$(\dot{\tilde{q}} + \lambda\tilde{q} + \dot{e} + \lambda e)^T M^{-1}(q)$$
$$\times [C(q, \dot{q}_d)\dot{q}_d - C(q, \dot{q})\dot{q} - F_v\dot{e}]$$
$$\leq (||\dot{\tilde{q}}|| + \lambda||\tilde{q}|| + ||\dot{e}|| + \lambda||e||)M_M C_M(||\dot{e}||^2$$
$$+2||\dot{q}_d||||\dot{e}|| + C_M^{-1}F_{v,M}||\dot{e}||)$$
$$\leq \beta(||\dot{\tilde{q}}||^2 + \lambda^2||\tilde{q}||^2 + 5||\dot{e}||^2 + \lambda^2||e||^2)$$
$$M_M C_M(||\dot{\tilde{q}}|| + \lambda||\tilde{q}|| + ||\dot{e}|| + \lambda||e||)||\dot{e}||^2 \qquad (6.60)$$

where the fact that for any real scalars a, b, and ϱ, $(\varrho a - \varrho^{-1}b)^2 = \varrho^2 a^2 + \varrho^{-2}b^2 - 2ab \geq 0$ is used.

$$(\dot{e} + \lambda e)^T K_d \dot{\tilde{q}}$$
$$\leq (||\dot{e}|| + \lambda||e||)K_{d,M}||\dot{\tilde{q}}||$$
$$\leq \frac{1}{2}K_{d,M}((1 + \varrho^{-2})||\dot{\tilde{q}}||^2 + \varrho^2||\dot{e}||^2 + \lambda^2||e||^2) \qquad (6.61)$$

where ϱ is a scalar. Because of the inequalities (6.59) through (6.61), there results (6.57).

It is concluded that there exist controller-observer gains K_p, K_d, L_p, L_d for a suitable choice of ϱ such that $\alpha_i > 0$, $i = 1,\ 2,\ 3,\ 4$. According to β-ball lemma, it follows that

$$\dot{L} \leq -\kappa_1 ||\dot{e}||^2 - \kappa_2 ||e||^2 - \kappa_3 ||\dot{\tilde{q}}||^2 - \kappa_4 ||\tilde{q}||^2 \tag{6.62}$$

where $\kappa_i > 0$, $i = 1,\ 2,\ 3,\ 4$. This allows us to write

$$\dot{L}(\dot{e}, e, \dot{\tilde{q}}, \tilde{q}) \leq -\kappa\ L(\dot{e}, e, \dot{\tilde{q}}, \tilde{q}) \tag{6.63}$$

for some $\kappa > 0$.

Therefore, the Lyapunov function, $L(\dot{e}, e, \dot{\tilde{q}}, \tilde{q})$, is negative definite. Thus, the closed-loop system, (6.51), (6.52), (6.46), is exponentially stable.

The drawback of this controller-observer scheme is that an exact knowledge about the robot system dynamics is required. In the following section, how to include an online neural estimator in the controller so that the need to know of system dynamics is minimized will be discussed.

6.5 Neurocontrol

Artificial neural networks (ANNs) are collections of mathematical models that try to emulate the observed properties of naturally occurring biological systems and draw on the analogies of adaptive biological learning. In the 1980s feedforward NNs were constructed and demonstrated to approximate quite well nearly all functions encountered in real-world applications. As a result of the work of numerous researchers, it came to be realized that NNs are capable of universal approximation in a very satisfactory sense. Owing to their universal approximation property, NNs have emerged as a powerful tool for controller design techniques for uncertain nonlinear systems including robot systems, which is a general class of nonlinear systems.

NNs have been applied in the field of control engineering recently and various results have been achieved. In the work of Narendra and Parthasarathy published in 1991 [18], NNs were proposed for the identification and control of nonlinear dynamical systems. As a result, a considerable number of papers have appeared on the general subject of control with NNs. In NN-based adaptive algorithms, since the desired or ideal weights are unknown, they are adapted real-time using suitably defined adaptation rules. Gradient methods were widely adopted [30]–[31]. Training schemes mainly based on the backpropagation algorithm [32] and its variants were proposed and applied to NN control. In recent contemporary literature, methods can be found to use NNs as online adaptive modules in closed-loop control with guaranteed stability. An NN in a closed-loop feedback control system also becomes a dynamical component part of the overall controller, and hence, the related issues of internal stability, weight convergence and robustness of the NN must be studied before conclusions about the closed-loop performance can be made.

Among the numerous mathematical models and architectures proposed for ANNs, radial-basis-function (RBF) NNs and multi-layer perceptron (MLP) NNs are the most popular.

6.5.1 Radial Basis Function (RBF) NNs

Radial basis function neural networks (RBF NNs) are popular for their simplicity, fast learning, and universal approximation properties. The construction of RBF NNs of Moody and Darken involves three different layers: *input layer* that consists of input nodes, *hidden layer* where each neuron computes its activation using a radial basis function, and *output layer* that builds a linear weighted sum of hidden layer activations to output the response of the network. The analytic expression of the activation of a more typical RBF networks is of product type given by

$$\psi_i(\boldsymbol{x}) = \prod_{j=1}^{q} \exp\left(-\frac{(x_j - c_{ij})^2}{2\sigma_{ij}^2}\right), \qquad i = 1, 2, \cdots, N \qquad (6.64)$$

where $\boldsymbol{x} = [x_1, x_2, \cdots, x_q]^T$ is the input vector. $\boldsymbol{c}_{ij} = [c_{i1}, c_{i2}, \cdots, c_{iq}]^T$ and $\boldsymbol{\sigma}_{ij} = [\sigma_{i1}, \sigma_{i2}, \cdots, \sigma_{iq}]^T$ are the center state and standard deviations of Gaussians associated with each element of input vector, respectively. N is the number of hidden-layer neurons.

Activation function of a sum type RBF network (also known as Gaussian sum nets) is given by

$$\psi_i(\boldsymbol{x}) = \sum_{j=1}^{q} \exp\left(-\frac{(x_j - c_{ij})^2}{2\sigma_{ij}^2}\right), \qquad i = 1, 2, \cdots, N. \qquad (6.65)$$

As seen in (6.64), the maximum activation of a product type RBF neuron is 1 (when $x_j = c_{ij} \; \forall \; j$). However, on the other hand, the maximum activation of (6.65) can be as large as q. As a result, the sensitivity of (6.65) (sum type RBF) is higher than that of (6.64) (product type RBF) for the same weight changes. To this end, sum type RBF can provide a faster convergence.

Let $\boldsymbol{\psi} \in \Re^N$ be the vector of activations then, the output of the NN, $\boldsymbol{y} \in \Re^p$, can be written as

$$\boldsymbol{y}(\boldsymbol{x}, \hat{W}) = \hat{W}^T \boldsymbol{\psi}(\boldsymbol{x}) \qquad (6.66)$$

where $\hat{W} \in \Re^{N \times p}$ is the connecting weight matrix.

According to the universal approximation, given a positive constant ε_N and a continuous function $f(\boldsymbol{x}) : C \to \Re^p$, where $C \subset \Re^q$ is a compact set, there exists a weight vector $\hat{W} = W$ such that the nonlinear function $f(\boldsymbol{x})$ can be approximated by the output $\boldsymbol{y}(\boldsymbol{x}, \hat{W})$ of the NN

$$f(\boldsymbol{x}) = \boldsymbol{y}(\boldsymbol{x}, W) + \boldsymbol{\varepsilon}(\boldsymbol{x}) \tag{6.67}$$

where $\boldsymbol{\varepsilon}(\boldsymbol{x}) \in \Re^p$ is the NN approximation error vector that satisfies $\|\boldsymbol{\varepsilon}(\boldsymbol{x})\| \le \varepsilon_N$. The ideal weights W are usually defined as those that minimize the supremum norm over C of $\boldsymbol{\varepsilon}$.

6.5.2 Multi-layer Perceptron (MLP) NNs

Multilayer NNs are composed of an input layer, an output layer, and at least one layer of nonlinear processing elements, which sum incoming signals and generate output signals according to some predefined function called the activation function. An m-layer network with the same activation function $\phi(\cdot)$ at each layer can be described by

$$\boldsymbol{y}(\boldsymbol{x}, \hat{W}) = \hat{W}_m^T \phi(\hat{W}_{m-1}^T \phi(...\phi(\boldsymbol{x}))) \tag{6.68}$$

where $\boldsymbol{y} \in \Re^p$ is the output vector, $\boldsymbol{x}[x_0, x_1, ..., x_q]^T$ the NN input vector with $x_0 = 1$. $\hat{W}_i \in \Re^{(N_i+1) \times N_{i+1}}; i = 1, ..., m$ are the weight matrix, which includes the threshold vector associated with the ith layer as its first column of $\hat{W}_i^T$, where $q = N_1, p = N_{m+1}$. The usual choice for the activation function is the sigmoidal function, defined as

$$1/(1 + e^{-\xi s})$$

where ξ is a constant and s is the sum of all incoming signals to the neuron.

According to the universal approximation property of MLP NNs, there exists a weight matrix, $\hat{W} = W$, such that the nonlinear function $f(\boldsymbol{x})$ can be approximated by the output $\boldsymbol{y}(\boldsymbol{x}, \hat{W})$ of the MLP NN architecture with m-layers satisfying $f(\boldsymbol{x}) = \boldsymbol{y}(\boldsymbol{x}, W) + \boldsymbol{\varepsilon}(\boldsymbol{x})$.

6.5.3 Identification-based Indirect Control

Identification-based indirect control method requires off-line preliminary training of the NN by the use of practical data. This method will be discussed taking the following example into discussion.

Example 6.4: The first autopilot systems were designed to perform some of the tasks of the pilot as continuous attention required from the pilots during flights of many hours led to serious fatigue. Nowadays, similar technologies are used in ground and surface vehicles and are also called autopilot systems. One example is self-steering gear available in ships, boats, and space crafts. An autopilot is a system used to guide a vehicle autonomously without assistance from a human being.

An autonomous vehicle following example [37] is discussed here. Ability of the artificial neural networks (NNs) to learn input-output relations through examples can be adopted to train them to emulate a human driver. In order to automate the steering and speed control of a vehicle while following a lead vehicle (autonomous vehicle following), NNs can be trained to capture a human driver's response (speed and steering angle commands) to different combinations of heading angle and range inputs. The vehicle that will be automated consists of a microprocessor servo system to implement steering, throttle, brakes, and transmission functions. A binocular stereo vision system onboard is used to obtain the range and heading angle of the lead vehicle. The block diagram of the sensing, processing, and actuation scheme of the vehicular system is given in Fig. 6.4. The driving command generator (DCG) that maps range and heading angle of the lead vehicle to control commands is to be realized using NNs. During the data collection

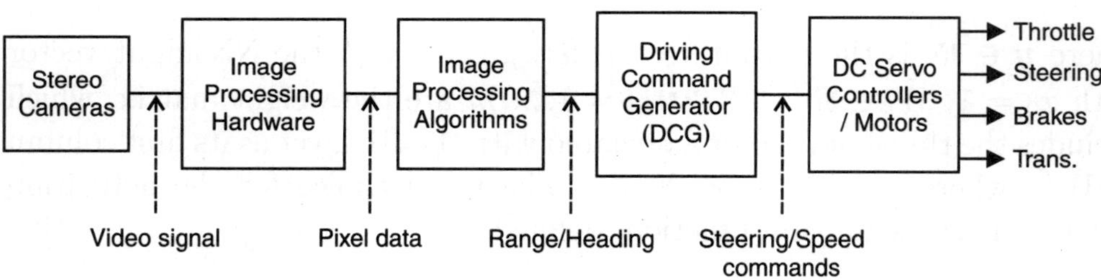

Figure 6.4: Block diagram of the autonomous vehicular system.

stage, the human driver drives the vehicle via a joystick (drive-by-wire). A drive-by-wire translator, binocular stereo camera system, and an image processor may also be used. During the data collection phase, the vehicle to be automated is driven by an expert human driver. These data will then be used to train NNs to capture the human driver response to different combinations of inputs that the driver responses to. The block diagram of manual driving set up during data collection is shown in Fig. 6.5.

The range and heading angle data from the vision system (inputs) and the commanded speed and steering wheel angle data from the drive-by-wire translator (outputs) are collected performing various maneuvers.

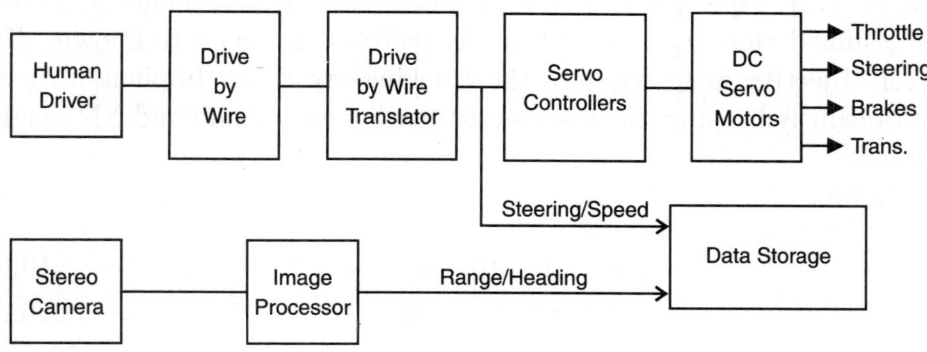

Figure 6.5: Block diagram of manual driving set up during data collection.

Two NNs, one for speed and the other for steering are then be trained off-line and used to reproduce human driver's driving. Training schemes for such off-line training mostly based on the backpropagation algorithm [32] and its variants.

The major drawback of such an identification-based indirect control approaches is that the controller may only be valid for operations under conditions prevailed during data collection. It may not yield the expected performance, for instance, on a rainy day when the roads are wet. To overcome such problems, online adaptive neural algorithms have been proposed and can be found in the literature.

6.5.4 Direct Closed-loop Neurocontrol

Example 6.4 described an example of implementation of an indirect approach of neurocontrol that is based on off-line training. However, methods that use NNs as online adaptive modules in closed-loop control with guaranteed stability are more theoretically sound. Ability to adapt may help deal with previously unknown situations. In such methods, the NN also becomes a dynamical component part of the overall controller, and hence, the related issues of internal stability, weight convergence and robustness of the NN must be studied through formative mathematical analyzes.

Design of a neurocontroller-observer scheme for robotic applications is described in this section. It is assumed that the inertia matrix, $M(\boldsymbol{q})$, of the rigid robot system is known. It is usual to have uncertainty in the Coriolis terms, which are difficult to compute, and the friction terms, which may have a complicated form.

In the sequel, $\hat{(\cdot)}$ denotes the estimated value of $(\cdot)$. Define $e = q - q_d$, and $\tilde{q} = q - \hat{q}$ where $q_d \in \Re^n$ is the desired joint angle trajectory. It is assumed that q_d and its time derivatives, $\dot{q}_d$, $\ddot{q}_d$, are known. The control objective is to regulate the tracking error, e. Implicit in it are simultaneously keeping the observer estimation error small and NN weights bounded.

Define

$$s = \dot{\tilde{q}} + \lambda \tilde{q} + \dot{e} + \lambda e \tag{6.69}$$

$$\hat{s} = \dot{\hat{q}} - \dot{q}_d + \lambda \tilde{q} + \lambda e \tag{6.70}$$

that yields

$$\hat{s} \equiv s - 2\dot{\tilde{q}}. \tag{6.71}$$

Consider the following combined controller-velocity observer system:
Controller:

$$\begin{cases} \tau(t) & = & M(q)\nu + \hat{W}^T \psi(x) - \epsilon\, \text{sgn}(\hat{s}^T M^{-1}(q))^T \\ \nu & = & \ddot{q}_d - K_d(\dot{\hat{q}} - \dot{q}_d) - K_p(q - q_d) \end{cases} \tag{6.72}$$

Observer:

$$\begin{cases} \dot{\hat{q}} & = & z + L_d(q - \hat{q}) \\ \dot{z} & = & \nu + L_p(q - \hat{q}) \end{cases} \tag{6.73}$$

where $\hat{W}^T \psi(x)$ is the NN functional estimate with x being the input vector, and ν is called a resolved acceleration in standard use. ϵ will be defined shortly. Matrices $K_d > 0$, $K_p > 0$ are assumed to be diagonal. $\dot{z} \in \Re^n$ denotes a reference acceleration input, which is obtained by modifying the resolved acceleration with position estimation error. Integrating $\dot{z}$ and further modifying it by position estimation error yield the estimated velocity.

The NN here is adopted as an online estimator to approximate the modified dynamic function, $C(q, \dot{q}_d)\dot{q}_d + G(q) + F_v\dot{q}_d + F_s(q)$. Input to the NN is hence chosen to be $x = [q^T \quad \dot{q}_d^T]^T$. Since the function is continuous, universal approximation property of RBF NNs, according to (6.66) and (6.67), allows us to write

$$C(q, \dot{q}_d)\dot{q}_d + G(q) + F_v\dot{q}_d + F_s(q) = W^T \psi(x) + \varepsilon(x) \tag{6.74}$$

where W is the matrix of ideal weights and $\|\varepsilon(x)\| \leq \varepsilon_N$.

Consider the controller-observer system defined by (6.72) and (6.73) in a closed loop with vehicle system (6.46). The closed-loop system can be made to be UUB by suitably selecting the controller-observer gains, K_p, K_d, L_p, and L_d.

Proof: In the sequel, we denote the minimum and maximum eigenvalue of any matrix $A(\boldsymbol{x}) = A^T(\boldsymbol{x}) > 0$ for all $\boldsymbol{x} \in \Re^n$ by A_m and A_M. Define $\tilde{W} = W - \hat{W}$. Consider the Lyapunov function candidate

$$L = \frac{1}{2}\dot{\tilde{\boldsymbol{q}}}^T\dot{\tilde{\boldsymbol{q}}} + \lambda\dot{\tilde{\boldsymbol{q}}}^T\tilde{\boldsymbol{q}} + \frac{1}{2}\tilde{\boldsymbol{q}}^T(L_p + \lambda L_d)\tilde{\boldsymbol{q}} + \frac{1}{2}\dot{\boldsymbol{e}}^T\dot{\boldsymbol{e}} + \lambda\dot{\boldsymbol{e}}^T\boldsymbol{e}$$

$$+ \frac{1}{2}\boldsymbol{e}^T(K_p + \lambda K_d)\boldsymbol{e} + \frac{1}{2}\text{tr}(\tilde{W}\Lambda^{-1}\tilde{W}^T) \tag{6.75}$$

which is positive definite for λ sufficiently small where $\Lambda = \Lambda^T > 0$ is a constant matrix and $\text{tr}(\cdot)$ is the trace of matrix $(\cdot)$. With (6.72), (6.73), (6.74), and (6.46), we obtain

$$M(\boldsymbol{q})(\ddot{\tilde{\boldsymbol{q}}} + L_d\dot{\tilde{\boldsymbol{q}}} + L_p\tilde{\boldsymbol{q}}) = C(\boldsymbol{q},\dot{\boldsymbol{q}}_d)\dot{\boldsymbol{q}}_d - C(\boldsymbol{q},\dot{\boldsymbol{q}})\dot{\boldsymbol{q}} - F_v\dot{\boldsymbol{e}}$$
$$+ \hat{W}^T\boldsymbol{\psi}(\boldsymbol{x}) - \epsilon\,\text{sgn}(\hat{s}^T M^{-1}(\boldsymbol{q}))^T$$
$$+ W^T\boldsymbol{\psi}(\boldsymbol{x}) + \boldsymbol{\varepsilon}(\boldsymbol{x}) \tag{6.76}$$

$$M(\boldsymbol{q})(\ddot{\boldsymbol{e}} + K_d\dot{\boldsymbol{e}} + K_p\boldsymbol{e} - K_d\dot{\tilde{\boldsymbol{q}}}) = C(\boldsymbol{q},\dot{\boldsymbol{q}}_d)\dot{\boldsymbol{q}}_d - C(\boldsymbol{q},\dot{\boldsymbol{q}})\dot{\boldsymbol{q}}$$
$$- F_v\dot{\boldsymbol{e}} + \hat{W}^T\boldsymbol{\psi}(\boldsymbol{x}) - \epsilon\,\text{sgn}(\hat{s}^T M^{-1}(\boldsymbol{q}))^T$$
$$+ W^T\boldsymbol{\psi}(\boldsymbol{x}) + \boldsymbol{\varepsilon}(\boldsymbol{x}). \tag{6.77}$$

Time differentiating (6.75), yields

$$\dot{L} = -\dot{\tilde{\boldsymbol{q}}}^T(L_d - \lambda I)\dot{\tilde{\boldsymbol{q}}} - \lambda\tilde{\boldsymbol{q}}^T L_p\tilde{\boldsymbol{q}} - \dot{\boldsymbol{e}}^T(K_d - \lambda I)\dot{\boldsymbol{e}}$$
$$- \lambda\boldsymbol{e}^T K_p\boldsymbol{e} + (\dot{\tilde{\boldsymbol{q}}} + \lambda\tilde{\boldsymbol{q}} + \dot{\boldsymbol{e}} + \lambda\boldsymbol{e})^T M^{-1}(\boldsymbol{q})$$
$$\times [C(\boldsymbol{q},\dot{\boldsymbol{q}}_d)\dot{\boldsymbol{q}}_d - C(\boldsymbol{q},\dot{\boldsymbol{q}})\dot{\boldsymbol{q}} - F_v\dot{\boldsymbol{e}}] + (\dot{\boldsymbol{e}} + \lambda\boldsymbol{e})^T K_d\dot{\tilde{\boldsymbol{q}}}$$
$$+ \boldsymbol{s}^T M^{-1}(\boldsymbol{q})[-\tilde{W}^T\boldsymbol{\psi}(\boldsymbol{x}) - \epsilon\,\text{sgn}(\hat{s}^T M^{-1}(\boldsymbol{q}))^T$$
$$- \boldsymbol{\varepsilon}(\boldsymbol{x})] + \text{tr}(\dot{\tilde{W}}\Lambda^{-1}\tilde{W}^T). \tag{6.78}$$

Select the weight tuning law as

$$\dot{\hat{W}} = -\boldsymbol{\psi}(\boldsymbol{x})\boldsymbol{s}^T M^{-1}(\boldsymbol{q})\Lambda + \kappa\left(W^0 - \hat{W}\right)\Lambda \tag{6.79}$$

where κ, W^0 are design parameters. It is shown that (6.78) can be upper bounded by

$$\dot{L} \leq -\alpha_2\|\boldsymbol{e}\|^2 - \alpha_3\|\dot{\tilde{\boldsymbol{q}}}\|^2 - \alpha_4\|\tilde{\boldsymbol{q}}\|^2$$
$$- (\alpha_1 - \gamma_1\|\dot{\tilde{\boldsymbol{q}}}\| - \gamma_2\|\tilde{\boldsymbol{q}}\| - \gamma_3\|\dot{\boldsymbol{e}}\| - \gamma_4\|\boldsymbol{e}\|)\|\dot{\boldsymbol{e}}\|^2$$
$$- \frac{1}{2}\kappa\|\tilde{W}\|^2 - \frac{1}{2}\kappa\|W^0 - \hat{W}\|^2 - \delta$$
$$+ 2\epsilon^2 M_M + \frac{1}{2}\kappa\|W^0 - W\|^2 \tag{6.80}$$

where

$$\alpha_1 = (K_{d,m} - \lambda - 5\beta - \frac{1}{2}\varrho^2 K_{d,M})$$

$$\alpha_2 = (\lambda K_{p,m} - \lambda^2\beta - \frac{1}{2}\lambda^2 K_{d,M})$$

$$\alpha_3 = (L_{d,m} - \lambda - \beta - \frac{1}{2}(1 - \varrho^{-2})K_{d,M} - 2M_M)$$

$$\alpha_4 = (\lambda_{p,m} - \lambda^2\beta)$$

$$\lambda\gamma_1 = \gamma_2 = \lambda\gamma_3 = \gamma_4 = \lambda M_M C_M. \tag{6.81}$$

Proof: Define $\beta = ||\dot{\boldsymbol{q}}_d|| M_M C_M + \frac{1}{2} M_M F_{v,M}$. Following inequalities can be written for the terms of (6.78)

$$-\dot{\tilde{\boldsymbol{q}}}^T(L_d - \lambda I)\dot{\tilde{\boldsymbol{q}}} - \lambda\tilde{\boldsymbol{q}}^T L_p\tilde{\boldsymbol{q}} - \dot{\boldsymbol{e}}^T(K_d - \lambda I)\dot{\boldsymbol{e}} - \lambda\boldsymbol{e}^T K_p\boldsymbol{e}$$
$$\leq -(L_{d,m} - \lambda)||\dot{\tilde{\boldsymbol{q}}}||^2 - \lambda L_{p,m}||\tilde{\boldsymbol{q}}||^2 - (K_{d,m} - \lambda)||\dot{\boldsymbol{e}}||^2$$
$$-\lambda K_{p,m}||\boldsymbol{e}||^2 \tag{6.82}$$

$$(\dot{\tilde{\boldsymbol{q}}} + \lambda\tilde{\boldsymbol{q}} + \dot{\boldsymbol{e}} + \lambda\boldsymbol{e})^T M^{-1}(\boldsymbol{q})$$
$$\times[C(\boldsymbol{q}, \dot{\boldsymbol{q}}_d)\dot{\boldsymbol{q}}_d - C(\boldsymbol{q}, \dot{\boldsymbol{q}})\dot{\boldsymbol{q}} - F_v\dot{\boldsymbol{e}}]$$
$$\leq (||\dot{\tilde{\boldsymbol{q}}}|| + \lambda||\tilde{\boldsymbol{q}}|| + ||\dot{\boldsymbol{e}}|| + \lambda||\boldsymbol{e}||)M_M C_M(||\dot{\boldsymbol{e}}||^2$$
$$+2||\dot{\boldsymbol{q}}_d||||\dot{\boldsymbol{e}}|| + C_M^{-1}F_{v,M}||\dot{\boldsymbol{e}}||)$$
$$\leq \beta(||\dot{\tilde{\boldsymbol{q}}}||^2 + \lambda^2||\tilde{\boldsymbol{q}}||^2 + 5||\dot{\boldsymbol{e}}||^2 + \lambda^2||\boldsymbol{e}||^2)$$
$$M_M C_M(||\dot{\tilde{\boldsymbol{q}}}|| + \lambda||\tilde{\boldsymbol{q}}|| + ||\dot{\boldsymbol{e}}|| + \lambda||\boldsymbol{e}||)||\dot{\boldsymbol{e}}||^2 \tag{6.83}$$

where the fact that for any real scalars a, b, and ϱ, $(\varrho a - \varrho^{-1}b)^2 = \varrho^2 a^2 + \varrho^{-2}b^2 - 2ab \geq 0$ is used.

$$(\dot{\boldsymbol{e}} + \lambda\boldsymbol{e})^T K_d\dot{\tilde{\boldsymbol{q}}}$$
$$\leq (||\dot{\boldsymbol{e}}|| + \lambda||\boldsymbol{e}||)K_{d,M}||\dot{\tilde{\boldsymbol{q}}}||$$
$$\leq \frac{1}{2}K_{d,M}((1 + \varrho^{-2})||\dot{\tilde{\boldsymbol{q}}}||^2 + \varrho^2||\dot{\boldsymbol{e}}||^2 + \lambda^2||\boldsymbol{e}||^2) \tag{6.84}$$

where ϱ is a scalar. Since $\tilde{W} = W - \hat{W}$ and W is constant, with the weight tuning law (6.79), there results

$$\mathrm{tr}(\dot{\tilde{W}}\Lambda^{-1}\tilde{W}^T) = -\mathrm{tr}(\dot{\hat{W}}\Lambda^{-1}\tilde{W}^T)$$
$$= -\mathrm{tr}\left(\left(-\boldsymbol{\psi}(\boldsymbol{x})\boldsymbol{s}^T M^{-1}(\boldsymbol{q}) + \kappa(W^0 - \hat{W})\right)\tilde{W}^T\right)$$

$$= s^T M^{-1}(q)\tilde{W}^T \psi(x) - \kappa \, \mathrm{tr}\left((W^0 - \hat{W})\tilde{W}^T\right)$$

$$\leq s^T M^{-1}(q)\tilde{W}^T \psi(x) - \frac{1}{2}\kappa\|\tilde{W}\|^2 - \frac{1}{2}\kappa\|W^0 - \hat{W}\|^2$$

$$+ \frac{1}{2}\kappa\|W^0 - W\|^2. \tag{6.85}$$

And, also

$$s^T M^{-1}(q)[-\tilde{W}^T \psi(x) - \epsilon \, \mathrm{sgn}(\hat{s}^T M^{-1}(q))^T - \varepsilon(x)]$$

$$+ \mathrm{tr}(\dot{\tilde{W}}\Lambda^{-1}\tilde{W}^T)$$

$$\leq s^T M^{-1}(q)[-\epsilon \, \mathrm{sgn}(\hat{s}^T M^{-1}(q))^T - \varepsilon(x)]$$

$$- \frac{1}{2}\kappa\|\tilde{W}\|^2 - \frac{1}{2}\kappa\|W^0 - \hat{W}\|^2 + \frac{1}{2}\kappa\|W^0 - W\|^2. \tag{6.86}$$

Now, select ϵ to the known upper bound defined as $\|\varepsilon(x)\| \leq \varepsilon_N \leq \epsilon$. Recalling $s = \hat{s} + 2\tilde{\dot{q}}$, we get

$$s^T M^{-1}(q)[-\epsilon \, \mathrm{sgn}(\hat{s}^T M^{-1}(q))^T - \varepsilon(x)]$$

$$\leq -\delta + 4\epsilon M_M\|\tilde{\dot{q}}\|$$

$$\leq -\delta + 2M_M\|\tilde{\dot{q}}\|^2 + 2\epsilon^2 M_M \tag{6.87}$$

where $\delta = \hat{s}^T M^{-1}(q)[-\epsilon \, \mathrm{sgn}(\hat{s}^T M^{-1}(q))^T - \varepsilon(x)] \leq 0$ according to the definition of ϵ. Because of the inequalities (6.82) through (6.87), there results (6.80).

It is concluded that there exist controller-observer gains K_p, K_d, L_p, L_d for a suitable choice of ϱ such that $\alpha_i > 0$, $i = 1, 2, 3, 4$. According to β-ball lemma, it follows that

$$\dot{L} \leq -\kappa_1\|\dot{e}\|^2 - \kappa_2\|e\|^2 - \kappa_3\|\tilde{\dot{q}}\|^2 - \kappa_4\|\tilde{q}\|^2$$

$$- \frac{1}{2}\kappa\|\tilde{W}\|^2 - \frac{1}{2}\kappa\|W^0 - \hat{W}\|^2 - \delta$$

$$+ 2\epsilon^2 M_M + \frac{1}{2}\kappa\|W^0 - W\|^2 \tag{6.88}$$

where $\kappa_i > 0$, $i = 1, 2, 3, 4$. This allowing us to write

$$\dot{L}(\dot{e}, e, \tilde{\dot{q}}, \tilde{q}, \tilde{W}) \leq -\frac{\vartheta}{P_M} \, L(\dot{e}, e, \tilde{\dot{q}}, \tilde{q}, \tilde{W}) + \gamma \tag{6.89}$$

where $\gamma = 2\epsilon^2 M_M + \frac{1}{2}\kappa\|W^0 - W\|^2$ and $\vartheta = 2\min\{\kappa_1, \kappa_2, \kappa_3, \kappa_4, \frac{1}{2}\kappa\}$. P_M is the positive constant that satisfies

$$L(\chi(t), \tilde{W}, t) \leq P_M(\|\chi(t)\|^2 + \|\tilde{W}\|_F^2) \tag{6.90}$$

where $\boldsymbol{\chi} = [\dot{\boldsymbol{e}}^T \ (\lambda \boldsymbol{e})^T \ \dot{\tilde{\boldsymbol{q}}}^T \ (\lambda \tilde{\boldsymbol{q}})^T]^T$ and $||\tilde{W}||_F = \sqrt{\operatorname{tr}(\tilde{W}\tilde{W}^T)}$.

Therefore, the Lyapunov function, $L(\dot{\boldsymbol{e}}, \boldsymbol{e}, \dot{\tilde{\boldsymbol{q}}}, \tilde{\boldsymbol{q}}, \tilde{W})$, is negative definite outside a compact set with the size of $O(\gamma)$ in the state space. Thus, the closed-loop system, (6.72), (6.73), (6.46), is uniformly ultimately bounded (UUB).

Remark 1: $\boldsymbol{s}$ in the weight adaptation law (6.79) is a function of unknown signal $\dot{\boldsymbol{q}}$. For weight update purpose alone, $\dot{\boldsymbol{q}}$ may be approximated using finite difference method as

$$\dot{\boldsymbol{q}}(k) = \frac{\boldsymbol{q}(k) - \boldsymbol{q}(k-1)}{T} \tag{6.91}$$

where k is the time index and T is the sampling interval.

6.5.5 Implementation

For a planar (vertical) two-link robot arm, for instance, the joint variable is $\boldsymbol{q} = [q_1 \ q_2]^T$. The dynamics of the robot arm are given below.

$$M(\boldsymbol{q}) = \begin{bmatrix} 9.77 + 2.02\cos(q_2) & 1.26 + 1.01\cos(q_2) \\ 1.26 + 1.01\cos(q_2) & 1.12 \end{bmatrix}$$

$$C(\boldsymbol{q}, \dot{\boldsymbol{q}}) = \begin{bmatrix} -1.01\sin(q_2)\dot{q}_2 & -1.01\sin(q_2)(\dot{q}_1 + \dot{q}_2) \\ 1.01\sin(q_2)\dot{q}_1 & 0 \end{bmatrix}$$

$$G(\boldsymbol{q}) = g \begin{bmatrix} 8.1\sin(q_1) + 1.13\sin(q_1 + q_2) \\ 1.13\sin(q_1 + q_2) \end{bmatrix}$$

where $g = 9.8\mathrm{ms}^{-2}$ is the acceleration of gravity. The friction terms are given by

$$F_v = [1.0 \ \ 1.2]^T, \qquad F_s(\boldsymbol{q}) = [10\cos(3q_1) \ \ 5\sin(2q_2)]^T.$$

Controller-observer scheme is implemented on the underlining robot systems as follows. Initial conditions are

$$\boldsymbol{q}(0) = [0.5 \ \ 3.0]^T, \qquad \hat{\boldsymbol{q}}(0) = \boldsymbol{q}(0)$$
$$\dot{\boldsymbol{q}}(0) = [0.0 \ \ 0.0]^T, \qquad \boldsymbol{z}(0) = \dot{\hat{\boldsymbol{q}}}(0) = \dot{\boldsymbol{q}}(0).$$

Controller-observer gains are selected as

$$K_d = \operatorname{diag}(50, \ 50), \ K_p = \operatorname{diag}(5000, \ 5000)$$
$$L_d = \operatorname{diag}(1, \ 1), \ L_p = \operatorname{diag}(0.1, \ 0.1).$$

And, $\lambda = 0.4$. The NN has $N = 10$ hidden neurons. Input vector to the neural network is $\boldsymbol{x} = [q_1 \; q_2 \; \dot{q}_{1d} \; \dot{q}_{2d}]^T$. Connecting weights are initialized in the range $[-0.5, 0.5]$, center states are set in the range $[-2, 2]$, and standard deviations are set at $3/\sqrt{2}$. Adaptation law parameters are selected as $\Lambda = \text{diag}(2000, \; 2000)$, $\kappa_x = \kappa_y = 0.01$, and $W^0 = \{w^0_{i,j} = 0.5\}_{n \times N}$.

Fig. 6.6 gives the desired and actual joint angle trajectories. The NNs functional estimates are depicted in Fig. 6.7.

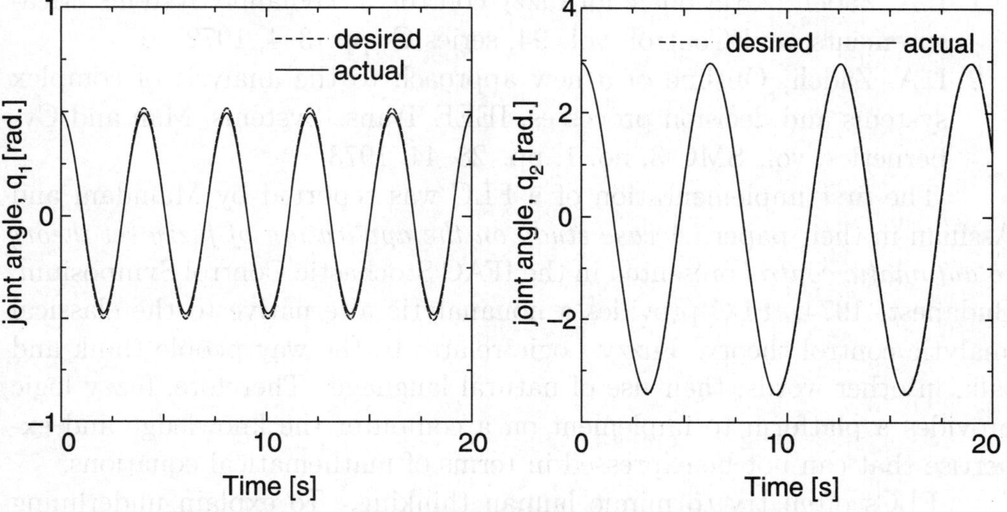

Figure 6.6: Joint angle trajectories-rigid robot arm.

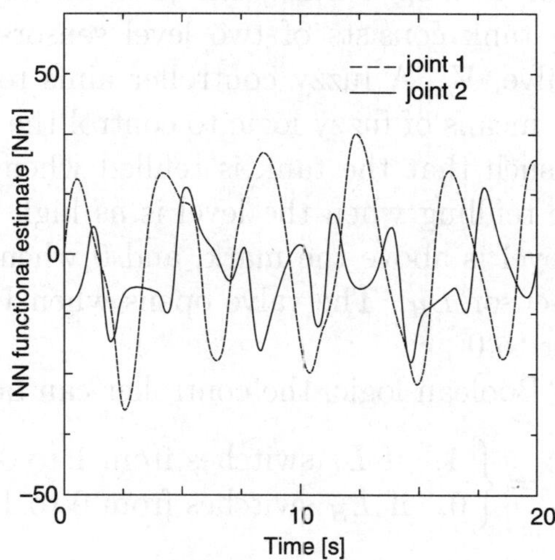

Figure 6.7: Neural network functional estimates.

6.6 Fuzzy Logic Control (FLC)

There are many areas of uncertainty or fuzziness in real world systems and an efficient way of dealing with this fuzziness is by the mechanism of fuzzy logic. Owing to its ease of implementation and robustness, fuzzy logic control (FLC) is increasingly growing in popularity among control engineers. The basic idea of FLC was suggested by Prof. L.A. Zadeh in his following publications.

1. L.A. Zadeh, A rationale for fuzzy control, J. Dynamic Systems, Measurements, and Control, vol. 94, series G, pp. 3–4, 1972

2. L.A. Zadeh, Outline of a new approach to the analysis of complex systems and decision processes, IEEE Trans. Systems, Man and Cybernetics, vol. SMC-3, no. 1, pp. 28–44, 1973.

The first implementation of a FLC was reported by Mamdani and Assilian in their paper, *A case study on the application of fuzzy set theory to automatic control* presented in the IFAC Stochastic Control Symposium, Budapest, 1974. FLC provides a nonanalytic alternative to the classical analytic control theory. Fuzzy Logic relates to the way people think and talk, in other words, their use of natural language. Therefore, fuzzy logic provides a platform to implement on a computer the knowledge and expertise that can not be expressed in terms of mathematical equations.

FLCs often try to mimic human thinking. To explain underlining fuzzy logic and to differentiate it from Boolean logic, consider a simple feed-tank that is for feeding a mill such that the feed flow is more or less constant. The tank consists of two level sensors, L_L and L_H, and a magnetic inlet valve, V. A fuzzy controller aims to mimic the human operator's terms by means of fuzzy logic to control the inlet valve, V. The valve is controlled such that the tank is refilled when the level is as low as L_L and stop the refilling when the level is as high as L_H. The sensor L_L is 1 when the level is above the mark, and 0 when the level is below. Likewise with the sensor L_H. The valve opens when V is set to 1, and it closes when V is set to 0.

In two-valued Boolean logic, the controller can be described as

$$V = \begin{cases} 1, & \text{if } L_L \text{ switches from 1 to 0} \\ 0, & \text{if } L_H \text{ switches from 0 to 1} \end{cases}$$

On the other hand, an human operator, whose responsibility is to open and close the valve, would perhaps describe the control strategy as

$$\text{If level is low, then open } V$$
$$\text{If level is high, then close } V.$$

The former strategy is suitable for a programmable logic controller (PLC) using Boolean logic, and the latter is suitable for a fuzzy controller (FLC) using fuzzy logic. Using FLC, one can even control the degree of valve opening as the level can be fed to FLC as a continuous input variable.

Fuzzy logic control is generated in a three-step process: fuzzification, inference, and defuzzification.

6.6.1 The Three-step Process of Generating FLCs

Think of an fuzzy logic controller (FLC) as the controller block in the control system schematic diagram. FLC receives inputs via the feedback loops and produces outputs to the plant. Outputs are produced according to the control strategy that is necessarily implemented using fuzzy logic. Inputs to FLC are typically signals that are available as measured deterministic or *crisp* signals.

In fuzzification, crisp inputs are mapped into membership functions of a fuzzy set. The membership function maps the crisp value to a membership value between 0 and 1. Each input has a fuzzy set that has one or many membership functions. For example, assume that a particular input ranges from -15 to +15. If -15 to 0 is considered *low* and 0 to +15 is considered *high*, then -7.5 to +7.5 may be considered *medium*. This input variable can be fuzzified or be mapped into membership functions as shown in Fig. 6.8.

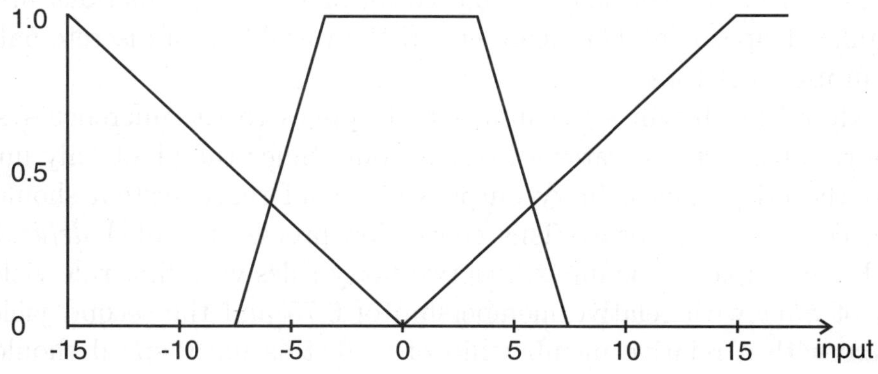

Figure 6.8: The relative membership in a fuzzy set.

A value of 3 of the input variable, for instance, has a membership of 1.0 in the function *Medium* and 0.2 in the function *High*. Therefore, this *degree* of membership notion in fuzzy logic, in contrast to Boolean logic, makes it suitable for continuous control applications. Before moving on to the *inference* stage, the outputs may also be fuzzified.

The control rules are developed in the inference stage of FLC design to relate fuzzified inputs to fuzzified outputs. Several rules can be written to define the control. For simplicity, assume that the controller has a single input and single output in a speed control problem of a motor. The rules can be

> *Rule 1* : If input is *Low*, then set motor speed to *Slow*
> *Rule 2* : If input is *Medium*, then set motor speed to *Medium*
> *Rule 3* : If input is *High*, then set motor speed to *Fast*

Here, *Low, Medium, High* are also called input fuzzy variables and *Slow, Medium, Fast* are output fuzzy variables. Typically, these rules take the form of IF-THEN statements. In FLCs having multiple inputs, the rules take the form If A AND B AND $C \cdots$, THEN O_1. It is noteworthy that since the rules are based on natural language, FLC is applicable to nonlinear systems and to control complex dynamical systems that do not have solid mathematical models. Fuzzy logic-based methods require an expert to write the linguistically expressible input-output relations or rules. Hence, such systems are also known as fuzzy expert systems. FLC design hence includes selection of the inputs and outputs, fuzzification, and inferencing. Note that fuzzification stage includes defining the range for each variable, deciding the number of membership functions, and their shapes and distribution. Variety of basic shapes are used for membership functions such as triangular, trapezoidal, Gaussian, etc. If the variable is of discrete nature, one can use singletons.

After FLC fuzzifies the inputs and applies to the inference system or the rule base, it generates a fuzzy output. Since the plant only understands the crisp values, fuzzy outputs of an inference system should be converted to a crisp form. This conversion process is called *defuzzification*. For example, if the input fires two fuzzy rules with first rule yielding speed of *Slow* with relative membership of 0.75 and the second yielding *Medium* with a relative membership of 0.50, this fuzzy speed should be defuzzified to obtain a crisp number such as 50 rpm. Defuzzification stage is straightforward as there are standard methods that can be used. Two common defuzzification methods are the *maximum defuzzificaiton method* and *centroid defuzzification method*. What centroid method produces is the centroid of the area under the curve. See Fig. 6.9. The crisp output is the velocity value corresponding to the centroid of the shaded area.

Matlab (www.mathworks.com) has a *fuzzy logic toolbox*. Combined with *Simulink*, this toolbox enables generate FLCs and perform computer

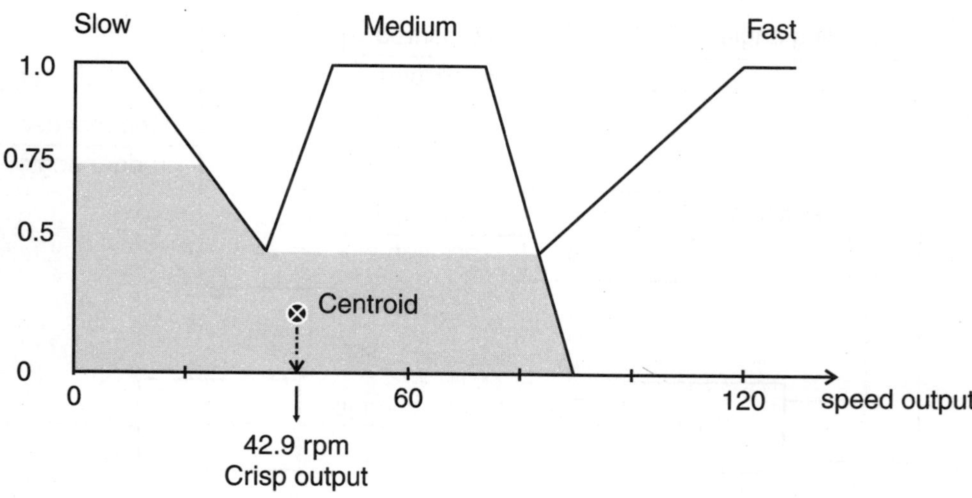

Figure 6.9: Centroidal defuzzification.

simulations. Parameters such as number of membership functions, their shape and distribution can be adjusted empirically to achieve the design specifications. FLCs works equally well with continuous time and discrete time systems.

6.6.2 Design Examples

Example 6.5: Fuzzy guidance controller can be designed to guide an autonomous boat [38]. The boat performs autonomous oceanographic research. It autonomously gather data to understand the topography of the sea floor. FLC is preferred because it is relatively easy to develop, it does not require complex mathematical model of the boat operating in the sea, and it is robust to external disturbances. Here, FLC is used to track and cross a series of predefined way-points. Way-points are defined as (x_i, y_i) in geographic coordinates and heading angle with reference to north.

The navigation data, position and heading, are obtained using GPS receiver (differential correction may be required) and compass, respectively. The control system block diagram is shown in Fig. 6.10.

In FLC design, intuition-based cause-and-effect relationship is used to define the inputs, fuzzify the inputs and the outputs, and to write the fuzzy rule base. One way-point is selected as the destination (target point) at a time. Once the current target point is reached, next way-point is taken as the target point or the new destination.

The inputs, $x(t)$, $y(t)$, and $\psi(t)$ are fuzzified as shown in Fig. 6.11 and

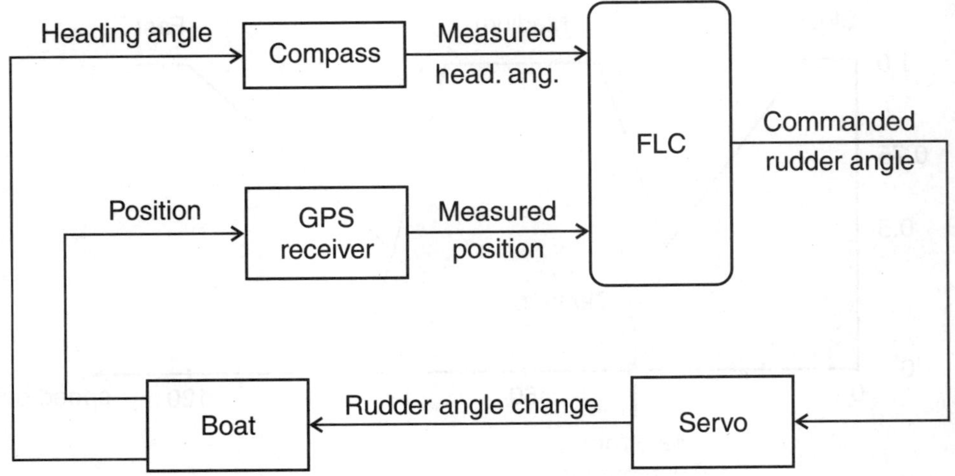

Figure 6.10: The control system block diagram for autonomous boat.

Fig. 6.12. The abbreviations used for labeling the membership functions are: N (negative), P (positive), B (big), M (medium), S (small), and ZE (zero).

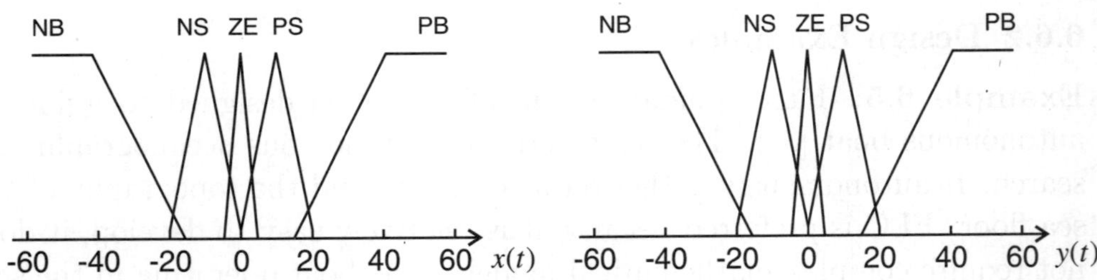

Figure 6.11: Membership functions of the position variables.

The output of the FLC is the commanded rudder angle, $\theta(t)$, where a positive rudder angle would produce a turn to the right. The output, $\theta(t)$, is fuzzified as shown in Fig. 6.13.

In the inferencing stage, fuzzy rules are written defining the mapping of the fuzzified inputs to the fuzzified outputs. Since 5, 5, and 7 membership functions are used to fuzzify $x(t)$, $y(t)$, and $\psi(t)$, respectively, the

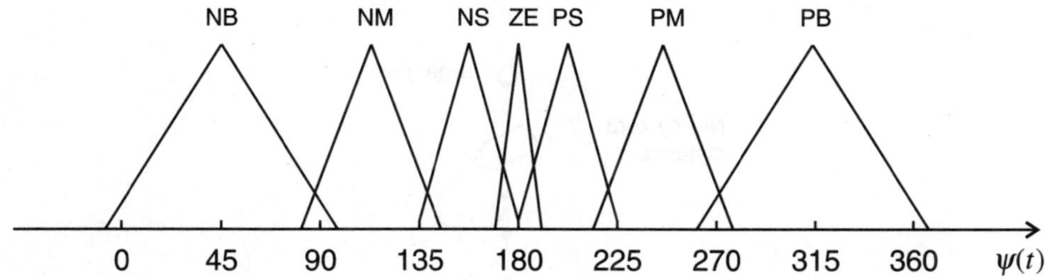

Figure 6.12: Membership functions for the heading variable. Desired value is 180°.

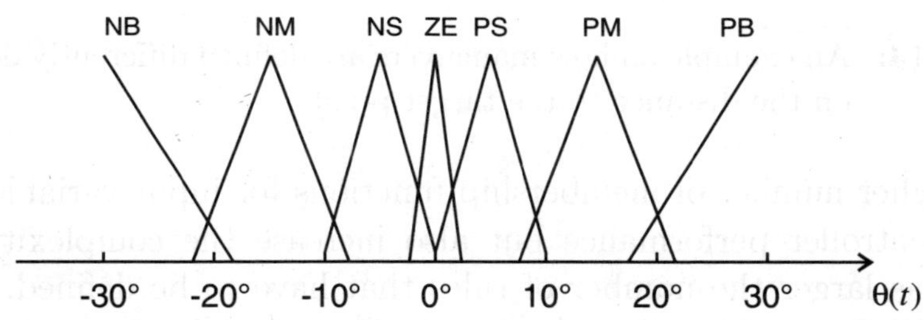

Figure 6.13: Membership functions for rudder angle.

inference engine will consist of a total of $5 \times 5 \times 7 = 175$ rules.

> *Rule 1* : If $(x \in NB)$ and $(y \in PB)$ and $(\psi \in NB)$ then $(\theta \in ZE)$
> *Rule 2* : If $(x \in NS)$ and $(y \in PB)$ and $(\psi \in NB)$ then $(\theta \in NM)$
> $\cdots$
> $\cdots$
> $\cdots$
> *Rule 175* : If $(x \in PB)$ and $(y \in NB)$ and $(\psi \in PB)$ then $(\theta \in NS)$

This rule bank is simply a mapping between measurements (position and heading) and action (steering command). They are defined in a way, for example, if the boat is far from the target point (a *large* distance) with enough time to maneuver, only a small steering correction is commanded. If the boat is closer (a *small* distance), then more aggressive command would turn the boat quickly. If the boat is very close (a *nearly zero* distance), and seemingly can not turn quickly enough to reach the target

point, a moderate turn to the opposite direction might be commanded to
give it time to recover. This is illustrated in Fig. 6.14.

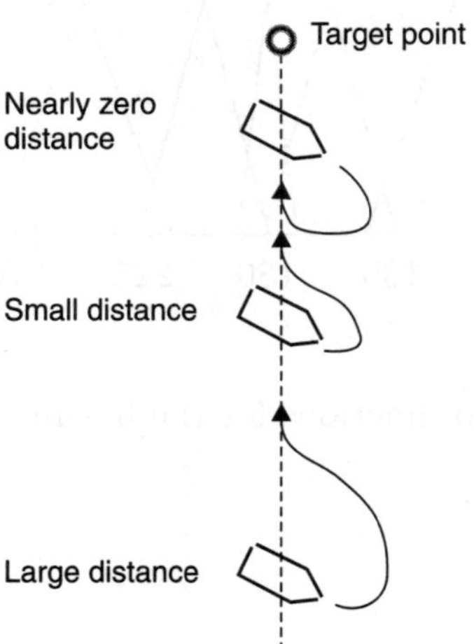

Figure 6.14: An example on how maneuvers are defined differently depending
on the distance to the target point.

A higher number of membership functions for input variables would
improve controller performance but also increase the complexity of the
controller as larger the number of rules that have to be defined. Part of
the art of FLC is making these trade-offs. One partial solution is to make
the membership functions near the desired value narrower to provide more
precise control when the system is operating close to the desired values.
This is practiced also with the output membership functions. In this case,
membership functions near zero rudder angle have been made narrower to
enable fine control when only a small steering correction is required.

Since hydrodynamic forces depends on the speed of the boat, its heading control may also be affected by the speed if the boat's speed varies in
a significant range. To take into account this coupling effect, speed should
also be included as an input variable making the input space 4-dimensional.
This complicates the FLC design as the number f rules to be defined will
at least be doubled. One important facet of design of control systems is to
design the controlled system in a way unnecessary complexities in the controller design phase are avoided. For instance, the rudder-behind-propeller
configuration of the boat can ensure that control effectiveness of the rudder does not depend significantly on the boat's speed. This simplifies the
controller design as adding speed as another input increases complexity of
the controller architecture. This is a good example for how the control
problem can be simplified by proper design of the controlled system.

Example 6.6: In people suffering from spinel cord injury, inability to move lower body is due to the loss of transmission of motor signals to the muscles. Despite this loss of communication between brain and the muscles, the muscles are still working and can be activated by artificial stimulaters. This is known as functional electrical stimulation (FES). Hence, a controller can be designed to generate suitable FES stimulating signals to perform rehabilitation cyclical exercises such as ergometric cycling and rowing.

Following fuzzy PD type FLC is used to stimulate the knee extensor muscles. The angle between the thigh and the shank is controlled to move the knee joint in a controlled manner. It is assumed that the angle feedback is available. The schematic of the control system is shown in Fig. 6.15.

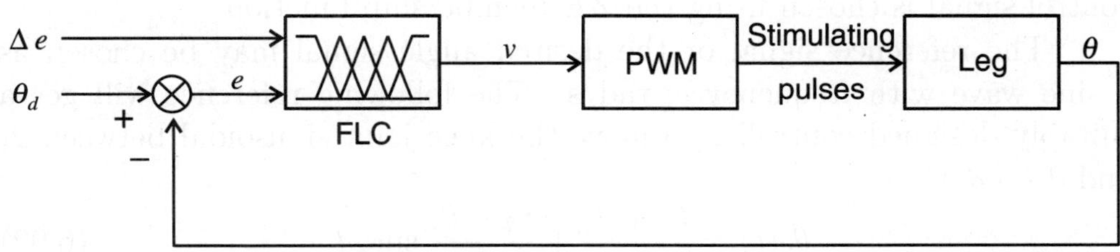

Figure 6.15: The control system block diagram for cyclical leg motion.

The error rate is approximated using finite difference approximation. Since the sampling time is constant, one may use $\Delta e = e(k) - e(k-1)$ and e as the two inputs where $e(t) = \theta_d(t) - \theta(t)$ with $\theta_d(t)$ and $\theta(t)$, respectively, being the desired angle and the measured angle. Both inputs may be fuzzified using 5 membership functions each. This will yield 25 rules.

$$Rule\ 1 :\ \text{If } (e \in NB) \text{ and } (\Delta e \in NB) \text{ then } (v \in NB)$$
$$Rule\ 2 :\ \text{If } (e \in NB) \text{ and } (\Delta e \in NS) \text{ then } (v \in NB)$$
$$\cdots$$
$$\cdots$$
$$\cdots$$
$$Rule\ 25 :\ \text{If } (e \in PB) \text{ and } (\Delta e \in PB) \text{ then } (v \in PB)$$

The complete rule base is given in Table 6.1.

The rule bank is simply a mapping between inputs (error and error rate) and the controller action (control signal to the PWM scheme). Hence, this is also an example for fuzzy logic-based implementation of PD control. The rules are defined in a way, for example, if the error (e) is negative and

Table 6.1: Fuzzy PD control rule base for cyclic leg motion

	$\Delta e \in NB$	NS	ZE	PS	PB
$e \in NB$	NB	NB	NS	ZE	ZE
NS	NB	NS	NS	ZE	ZE
ZE	NB	NS	ZE	PS	PB
PS	ZE	ZE	PS	PS	PB
PB	ZE	ZE	PB	PB	PB

$\Delta e = e(k) - e(k-1)$ is positive, i.e., e is moving towards zero, then the control signal is chosen using the ZE membership function.

The reference signal or the desired angle signal may be chosen as a sine wave with frequency ω rad/s. The following reference will get a suitably designed controller to move the knee joint sinusoidal between θ_1 and $\theta_2(> \theta_1)$.

$$\theta_d(t) = \frac{\theta_1 + \theta_2}{2} + \frac{\theta_2 - \theta_1}{2} \sin \omega t. \tag{6.92}$$

For instance, select $\theta_1 = 80°$ and $\theta_2 = 100°$ to move the knee joint sinusoidal between $80°$ and $100°$. Speed can be adjusted by adjusting ω and the swing by adjusting α in $\theta_2 \to \theta_2 + \alpha$, $\theta_1 \to \theta_1 - \alpha$. To bring the leg motion to stop gradually, $\theta_d(t)$ may be changed to

$$\theta_{d_{stop}}(t) = e^{-at}\theta_d(t) \tag{6.93}$$

where $a > 0$. The user can make these adjustments and changes by the use of a simple hand-held interface device.

6.7 Neuro-fuzzy Systems

Fig. 6.16 shows an adaptive neuro-fuzzy inference system (ANFIS) architecture of linear Takagi-Sugeno type that can be used for synthesizing controllers [39], [40]. ANFIS controllers are very effective neuro-fuzzy *hybrid* schemes based on a fuzzy inference system framework with adaptive network functionality. The main advantage of using an ANFIS is that it enhances fuzzy controllers with self-learning capability for achieving the control objectives with near optimality.

ANFIS is a multilayer feed-forward network as shown in Fig. 6.16. Here, each node performs a particular function on incoming signals. That functional aspect is characterized with a set of parameters pertaining to

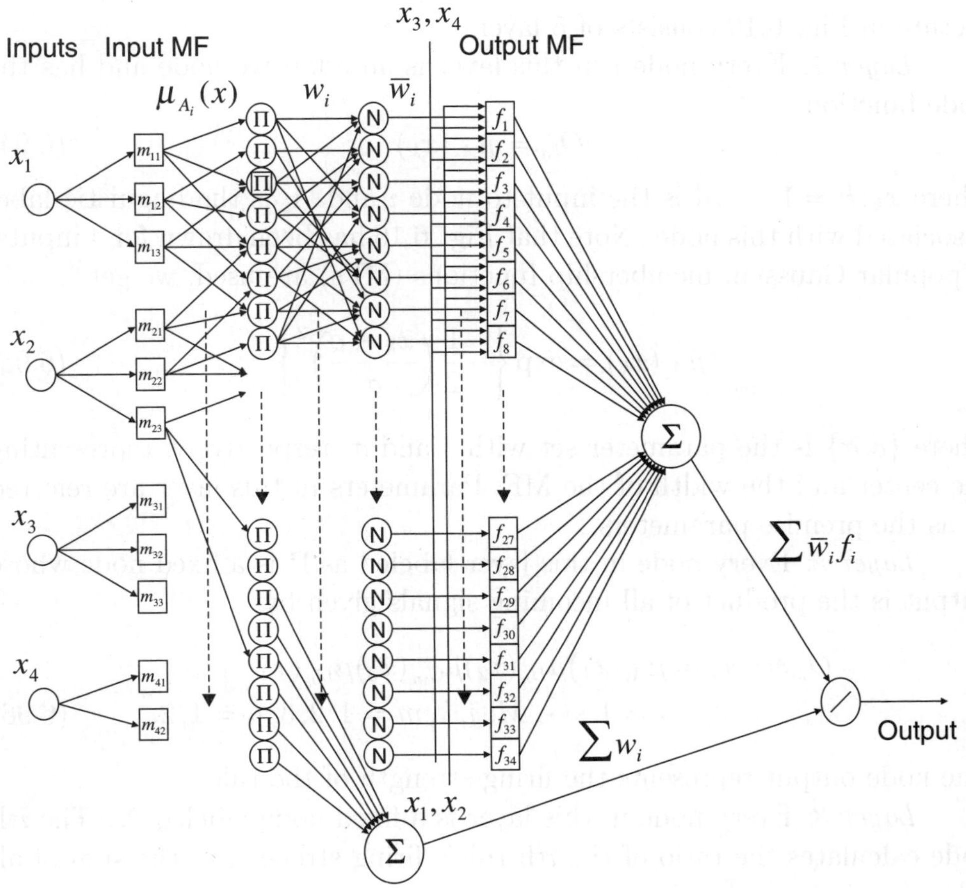

Figure 6.16: ANFIS architecture.

each node. The square nodes, which are commonly referred to as adaptive nodes, have parameters while the circular nodes (fixed nodes) have none. The collection of adaptive parameters is referred to as the parameter set of the adaptive network. To achieve the desired input-output mapping, these parameters are updated with the given training data.

ANFIS can be trained with popular back-propagation method. Other methods include the effective hybrid learning rule. With this learning algorithm, node outputs go forward until level 4 and the consequent parameters are identified by the least squares method. In the backward pass, the error signals propagate backwards and premise parameters are updated by gradient decent. This hybrid approach converges much faster than the original pure back-propagation method because the latter reduces the search space dimensions.

Let the output of the ith node in layer l be $O_{l,i}$. The ANFIS architecture in Fig. 6.16 consists of 5 layers

Layer 1: Every node i in this layer is an adaptive node and has the node function

$$O_{l,i} = \mu_{A_i}(x_k) \tag{6.94}$$

where $x_k; k = 1, \cdots, 4$ is the input to node i and A_i is the linguistic label associated with this node. Note that Fig. 6.16 has been drawn for 4 inputs. If popular Gaussian membership functions (MFs) are used, we get

$$\mu_{A_i}(x_k) = \exp\left\{ -\frac{1}{2}\left(\frac{x_k - c}{\sigma}\right)^2 \right\} \tag{6.95}$$

where $\{c, \sigma\}$ is the parameter set with c and σ, respectively, representing the center and the width of the MF. Parameters in this layer are referred to as the premise parameters.

Layer 2: Every node in this layer labeled as Π is a fixed node whose output is the product of all incoming signals given by

$$O_{2,i} = w_i = \mu_{A_i}(x_1)\mu_{B_k}(x_2)\mu_{C_m}(x_3)\mu_{D_n}(x_4)$$
$$i = 1, \cdots, M; j, k, m = 1, 2, 3; n = 1, 2. \tag{6.96}$$

The node output represents the firing strength of the rule.

Layer 3: Every node in this layer is a fixed node labeled N. The ith node calculates the ratio of the ith rule's firing strength to the sum of all rules' firing strengths given by

$$O_{3,i} = \bar{w}_i = \frac{w_i}{\sum w_i}; i = 1, 2, \cdots, M. \tag{6.97}$$

The outputs of this layer are the normalized firing strengths.

Layer 4: Every node i in this layer is an adaptive node with a node function

$$O_{4,i} = \hat{w}_i f_i = \hat{w}_i(p_i x_1 + q_i x_2 + r_i x_3 + s_i x_4 + t_i) \tag{6.98}$$

where $\hat{w}_i$ is the normalized firing strength from layer 3, $i = 1, 2, \cdots, M$, and $\{p_i, q_i, r_i, s_i, t_i\}$ is the parameter set of this node. Parameters in this layer are referred to as consequent parameters.

Layer 5: The single node in this layer is a fixed node labeled Σ, which computes the overall output as the sum of all incoming signals given by

$$O_{5,i} = \sum \bar{w}_i f_i = \frac{\sum w_i f_i}{\sum w_i} \tag{6.99}$$

The ANFIS architecture presented here has been used for steering control of a smart vehicle in [39] and has 4 input variables and one output. The rule base contains M number of fuzzy rules.

References

[1] Sisil Kumarawadu, *Modeling and Control of Vehicular and Robotic Systems*, Narosa Publishing House, Alpha Sci. Intl. Ltd., 2008.

[2] Ogata Katsuhiko, *Discrete-Time Control Systems - 2nd Edition*, Pearson Education Asia, 2002.

[3] N. Mohan, T.M. Undeland, and W.P. Robbins *Power Electronics - 3rd Edition*, John Willey & Sons, Inc., 2003.

[4] F.L. Lewis, C.T. Abdallah, and D.M. Dawson, *Control of Robot Manipulators*. New York: McMillan, 1993.

[5] Craig, J.J., *Introduction to Robotics: Mechanics and Control - 2nd Edition*, Reading, MA: Addision-Wesley, 1989.

[6] *2005 DSP56000 24-Bit Digital Signal Processor Family Manual*, Freescale Semiconductor, Inc., 2005.

[7] *DSP56005: 24-Bit Digital Signal Processor*, Motorola Semiconductor technical Data, Freescale Semiconductor, Inc., 1996.

[8] *DS1104 R&D Controller Board: A Powerful Prototyping System in Your PC*, Product Bulletin, dSPACE GmbH, 2001.

[9] *PCI-8392/PCI-8392H, DSP-based SSCNET III 16-axis Motion Controller*, Datasheet, ADLINK Technology, Inc., 2009.

[10] Aengus Murray, "Motion Control Chip-sets,", Analog Dialog, Analog Devices, Vol. 30, no. 2, pp. 3–6, 1996.

[11] J. Xiao, A. Calle, J. Ye, and Z. Zhu, "A Mobile Robot Platform with DSP-based Controller and Omnidirectional Vision System," in *Proc. of 2004 IEEE International Conference on Robotics and Biomimetics*, Shenyang, China, pp. 844–848, 2004.

[12] *F.A.A.K. Mobile Robot Project*: http://prt.fernuni-hagen.de/pro/faak

[13] dsPIC30F6010A/6015 Data Sheet, Microchip Technology Inc., 2005.

[14] Vinaya Skanda, "Power Factor Correction in Power Conversion Applications Using the dsPIC DSC,", Application note, Microchip Technology Inc., 2007.

[15] Horn, B.K.P., *Robot Vision*, Cambridge, MA: MIT Press, 1986.

[16] H. Berghuis, P. Löhnberg, and H. Nijmeijer, "Tracking control of robots using only position measurements," in *Proc. of IEEE 30th*

Conf. on Decision and Control, Brighton, England, pp. 1039–1040, 1991.

[17] J.T. Wen and D.S. Bayaerd, "New Class of Control Laws for Robotic Manipulator: Non-adaptive Case," *Int. Journal of Control*, Vol. 47, pp. 1361–1361, 1988.

[18] Narendra, K.S. and K. Parthasarathy., "Identification and Control of Dynamical Systems Using Neural Networks," *IEEE Trans. on Neural Networks*, Vol. 1, no. 1, pp. 4–27, 1990.

[19] Hornic, K., M. Stinchcombe, and H. White, "Universal Approximation of an Unknown Mapping and its Derivatives using Multilayer Feedforward Networks," *Neural Networks*, Vol. 3, pp. 551–560, 1990.

[20] Seshagiri, S. and H. Khalil, "Output Feedback Control of Nonlinear Systems using RBF Neural Networks," *IEEE Trans. on Neural Networks*, Vol. 11, no. 1, pp. 69–79, 2000.

[21] Lewis, F.L., A. Yesildirek and K. Liu, "Multilayer Neural-net Robot Controller with Guaranteed Tracking Performance," *IEEE Trans. on Neural Networks*, Vol. 7, no. 2, pp. 388–399, 1996.

[22] Lewis, F.L., K. Liu and A. Yesildirek, "Neural Net Robot Controller with Guaranteed Tracking Performance," *IEEE Trans. on Neural Networks*, Vol. 6, no. 3, pp. 703–715, 1995.

[23] Moody, J.E. and C. Darken, "Fast Learning in Networks of Locally-tuned Processing Units," *Neural Computation*, Vol. 1, pp. 290–294, 1989.

[24] Wang, L.X. and J.M. Mendel, "Fuzzy Basis Functions, Universal Approximation, and Orthogonal Least-squares Learning," *IEEE Trans. on Neural Networks*, Vol. 3, pp. 807–814, 1992.

[25] Lowe. D., *Adaptive Fuzzy Systems and Control*, Englewood Cliffs, NJ: Prentice-Hall, 1994.

[26] Ertugrul, M. and O. Kaynak, "Neuro-sliding Mode Control of robotic Manipulators," in *Procs. of The 8th International Conference on Advanced Robotics*, Monterey, CA, USA: pp. 951–956, 1997.

[27] Lee, B.K., and L.W. Mau, "An Output Tracking VSS Control in the Presence of a Class of Mismatched Uncertainties," *Asian Journal of Control*, Vol. 4, no. 2, pp. 206–216, 2002.

[28] Sisil Kumarawadu, K. Watanabe, K. Kiguchi, and K. Izumi, "An Active Binocular Vision Head, its Kinematic Analysis and Derivation of Equations of Motion," *JSME Int. J. Ser. C*, Vol. 46, no. 2, pp. 754-765, Jun. 2003.

[29] Sisil Kumarawadu, Keigo Watanabe, and Tsu-Tian Lee, "High-Performance Binocular Tracking with an Online Neural Estimator," *IEEE Transactions on Systems, Man and Cybernetics–Part B: Cybernetics*, Vol. 37, no. 1, pp. 213–223, Feb. 2007.

[30] Song, Q., J. Xiao and Y.C. Soh, "Robust Backpropagation Training Algorithm for Multilayered Neural Tracking controller," *IEEE Trans. on Neural Networks*, Vol. 10, no. 5, pp. 1133–1141, 1999.

[31] Tzafestas, E.S., A. Nikolaidou and S.G. Tzafestas, "Performance evaluation, and dynamic node-generation Criteria for "Principal Component Analysis" Neural Networks," *Mathematics and Computers in Simulation*, Vol. 51, no. 3–4, pp. 145–156, 2000.

[32] Werbos, P.J., "Beyond Regression: New Tools for Prediction and Analysis in the Behavioral Sciences," Ph.D. Dissertation, Harvard Univ., Boston, M.A., 1974.

[33] Tzafestas, S.G., P.J. Dalianis and A. Anthopoulos, "On the Overtraining Phenomenon of Back Propagation Neural Networks," *Mathematics and Computers in Simulation*, Vol. 40, nos. 5–6, pp. 507–521, 1996.

[34] J. Park and I.W. Sandberg, "Universal Approximation using Radial-basis-function Networks," *Neural Computation*, Vol. 3, pp. 246–257, 1991.

[35] H. Berghuis and H. Nijmeija, "A Passivity Approach to Controller-Observer Design for Robots," *IEEE Trans. Robotics and Automation*, Vol. 9, no. 6, pp. 740–754, 1993.

[36] F.L. Lewis, A. Yeşildirec, and K. Liu, "Neural-net Robot Controller with Guaranteed Tracking Performance," *IEEE Trans. Neural Networks*, Vol. 6, no. 3, pp. 703–715, 1995.

[37] N. Kehtarnavaz, N. Griswold, K. Miller, and P. Lescoe, "A Transportable neural-Network Approach to Autonomous Vehicle Following," *IEEE Trans. Vehicular Technology*, Vol. 47, no. 2, pp. 694–702, 1998.

[38] T.W. Vaneck, "Fuzzy Guidance Controller for an Autonomous Boat," *IEEE Control Systems*, Vol.17, pp. 44–47, 1997.

[39] S.R. Ranatunga and Sisil Kumarawadu, "Cooperatively Controlled Collision Evasive Emergency Manoevers," *Control and Intelligent Systems*, ACTA Press/IASTED, Vol. 36, no. 4, pp. 330–339, 2008.

[40] J. Mar and F.J. Lin, "An ANFIS Controller for the Car-following Collision Prevention System," *IEEE Trans. Vehicular Technology*, Vol. 50, no. 4, pp. 1106–1113, 2001.